THE INTERNATIONAL SCIENTIFIC SERIES.

VOLUME VI.

THE INTERNATIONAL SCIENTIFIC SERIES.

Works already Published.

I. FORMS OF WATER, IN CLOUDS, RAIN, RIVERS, ICE, AND GLACIERS. By Prof. JOHN TYNDALL, LL. D., F. R. S. 1 vol. Cloth. Price, $1.50.

II. PHYSICS AND POLITICS; OR, THOUGHTS ON THE APPLICATION OF THE PRINCIPLES OF "NATURAL SELECTION" AND "INHERITANCE" TO POLITICAL SOCIETY. By WALTER BAGEHOT, Esq., author of "The English Constitution." 1 vol. Cloth. Price, $1.50.

III. FOODS. By EDWARD SMITH, M. D., LL. B., F. R. S. 1 vol. Cloth. Price, $1.75.

IV. MIND AND BODY: THE THEORIES OF THEIR RELATIONS. By ALEX. BAIN, LL. D., Professor of Logic in the University of Aberdeen. 1 vol., 12mo. Cloth. Price, $1.50.

V. THE STUDY OF SOCIOLOGY. By HERBERT SPENCER. Price, $1.50.

VI. THE NEW CHEMISTRY. By Prof. JOSIAH P. COOKE, Jr., of Harvard University. 1 vol., 12mo. Cloth. Price, $2.00.

VII. THE CONSERVATION OF ENERGY. By Prof. BALFOUR STEWART, LL. D., F. R. S. 1 vol., 12mo. Cloth. Price, $1.50.

VIII. ANIMAL LOCOMOTION; OR, WALKING, SWIMMING, AND FLYING, WITH A DISSERTATION ON AËRONAUTICS. By J. BELL PETTIGREW, M. D., F. R. S. E., F. R. C. P. E. 1 vol., 12mo. Fully illustrated. Price, $1.75.

IX. RESPONSIBILITY IN MENTAL DISEASE. By HENRY MAUDSLEY, M. D. 1 vol., 12mo. Cloth. Price, $1.50.

X. THE SCIENCE OF LAW. By Prof. SHELDON AMOS. 1 vol., 12mo. Cloth. Price, $1.75.

XI. ANIMAL MECHANISM. A TREATISE ON TERRESTRIAL AND AËRIAL LOCOMOTION. By E. J. MAREY. With 117 Illustrations. Price, $1.75.

XII. THE HISTORY OF THE CONFLICT BETWEEN RELIGION AND SCIENCE. By JOHN WM. DRAPER, M. D., LL. D., author of "The Intellectual Development of Europe." Price, $1.75.

XIII. THE DOCTRINE OF DESCENT, AND DARWINISM. By Prof. OSCAR SCHMIDT, Strasburg University. Price, $1.50.

XIV. THE CHEMISTRY OF LIGHT AND PHOTOGRAPHY. IN ITS APPLICATION TO ART, SCIENCE, AND INDUSTRY. By Dr. HERMANN VOGEL. 100 Illustrations. Price, $2.00.

XV. FUNGI; THEIR NATURE, INFLUENCE, AND USES. By M. C. COOKE, M. A., LL. D. Edited by Rev. M. J. BERKELEY, M. A., F. L. S. With 109 Illustrations. Price, $1.50.

XVI. THE LIFE AND GROWTH OF LANGUAGE. By Prof. W. D. WHITNEY, of Yale College. Price, $1.50.

XVII. MONEY AND THE MECHANISM OF EXCHANGE. By W. STANLEY JEVONS, M. A., F. R. S., Professor of Logic and Political Economy in the Owens College, Manchester. Price, $1.75.

XVIII. ANIMAL PARASITES AND MESSMATES. By Monsieur VAN BENEDEN, Professor of the University of Louvain, Correspondent of the Institute of France. With 83 Illustrations. (*In press.*)

THE INTERNATIONAL SCIENTIFIC SERIES.

THE NEW CHEMISTRY.

BY

JOSIAH P. COOKE, Jr.,

ERVING PROFESSOR OF CHEMISTRY AND MINERALOGY IN HARVARD UNIVERSITY.

NEW YORK:

D. APPLETON AND COMPANY,

549 & 551 BROADWAY.

1876.

TO

JOHN A. LOWELL, LL.D.,

BOSTON, MASSACHUSETTS.

DEAR SIR:

From the early lectures of the Lowell Institute I derived, when a boy, my taste for the science which became the occupation of my after-life, and it has since often been my privilege to illustrate before the intelligent audiences—which, for more than thirty winters, the Institute has gathered under your direction—the results of the studies that I there began. Allow me, then, to dedicate to you this volume, as an expression of my indebtedness to the foundation you have so long and so ably administered.

With great respect,

Your obedient servant,

JOSIAH P. COOKE, JR.

PREFACE.

The lectures now published were delivered before the Lowell Institute, in Boston, in the autumn of 1872. They aimed to present the modern theories of chemistry to an intelligent but not a professional audience, and to give to the philosophy of the science a logical consistency, by resting it on the law of Avogadro. Since many of the audience had studied the elements of chemistry, as they were formerly taught under the dualistic system, it was also made an object to point out the chief characteristics by which the new chemistry differed from the old. The limitations of a course of popular lectures necessarily precluded a full presentation of the subject, and only the more prominent and less technical features of the new system were discussed. In writing out his notes for the press, the author has retained the lecture style, because it is so well adapted for the popular exposition of scientific subjects; but he is painfully conscious that any description of experiments must necessarily fall far short of giving that force of impression which the phenomena of Nature produce when they speak for themselves, and, in weighing the arguments presented, he must beg his readers to make allowances for this fact.

CAMBRIDGE, *September* 6, 1873.

CONTENTS.

THE NEW CHEMISTRY.

LECTURE I.

MOLECULES AND AVOGADRO'S LAW.

In every physical science we have carefully to distinguish between the facts which form its subject-matter and the theories by which we attempt to explain these facts, and group them in our scientific systems. The first alone can be regarded as absolute knowledge, and such knowledge is immutable, except in so far as subsequent observation may correct previous error. The last are, at best, only guesses at truth, and, even in their highest development, are subject to limitations, and liable to change.

But this distinction, so obvious when stated, is often overlooked in our scientific text-books, and not without reason, for it is the sole aim of these elementary treatises to teach the present state of knowledge, and they might fail in their object if they attempted, by a too critical analysis, to separate the phenomena from the systems by which alone the facts of Nature are correlated and rendered intelligible.

When, however, we come to study the history of science, the distinction between fact and theory obtrudes itself at once upon our attention. We see that, while the prominent facts of science have remained the same, its history has been marked by very frequent revolutions in its theories or systems. The courses of the planets have not changed since they were watched by the Chaldean astronomers, three thousand years ago; but how differently have their motions been explained—first by Hipparchus and Ptolemy, then by Copernicus and Kepler, and lastly by Newton and Laplace!—and, however great our faith in the law of universal gravitation, it is difficult to believe that even this grand generalization is the final result of astronomical science.

Let me not, however, be understood to imply a belief that man cannot attain to any absolute scientific truth; for I believe that he can, and I feel that every great generalization brings him a step nearer to the promised goal. Moreover, I sympathize with that beautiful idea of Oersted, which he expressed in the now familiar phrase, "*The laws of Nature are the thoughts of God;*" but, then, I also know that our knowledge of these laws is as yet very imperfect, and that our human systems must be at the best but very partial expressions of the truth. Still, it is a fact, worthy of our profound attention, that in each of the physical sciences, as in astronomy, the successive great generalizations which have marked its progress have included and expanded rather than superseded those which went before them. Through the great revolutions which have taken place in the forms of thought, the elements of truth in the successive systems have been preserved, while the error has been as constantly eliminated; and so, as I believe, it always will be, until the last generalization of all brings us into the presence of that law which is indeed the thought of God.

There is also another fact, which has an important

bearing on the subject we are considering. Almost all the great generalizations of science have been more or less fully anticipated, at least in so far that the general truth which they involve has been previously conceived. The Copernican theory was taught, substantially, by the disciples of Pythagoras. The law of gravitation was suggested, both by Hooke and Cassini, several years before Newton published his "Principia;" and the same general fact has been recently very markedly illustrated in the discovery of the methods of spectrum analysis, every principle of which had been previously announced. The history of science shows that the age must be prepared before really new scientific truths can take root and grow. The barren premonitions of science have been barren because these seeds of truth fell upon unfruitful soil; and, as soon as the fullness of the time was come, the seed has taken root and the fruit has ripened. No one can doubt, for example, that the law of gravitation would have been discovered before the close of the seventeenth century if Newton had not lived; and it is equally true that, had Newton lived before Galileo and Kepler, he never could have mastered the difficult problems it was his privilege to solve. We justly honor with the greatest veneration the true men who, having been called to occupy these distinguished places in the history of science, have been equal to their position, and have acquitted themselves so nobly before the world; but every student is surprised to find how very little is the share of new truth which even the greatest genius has added to the previous stock. Science is a growth of time, and, though man's cultivation of the field is an essential condition of that growth, the development steadily progresses, independently of any individual investigator, however great his mental power. The greatest philosophical generalizations, if premature, will fall on barren soil, and, when the age is ripe, they are never long delayed. The very discovery of law is regulated by law, or, as we rather believe, is directed by Providence; but, however we may prefer to represent the facts, this natural growth of knowledge gives us the strongest assurance that the growth is sound and the progress real. Although the foundations of science have been laid in such obscurity, its students have worked under the direction of the same guiding power which rules over the whole of Nature, and it cannot be that the structure they have reared with so much care is nothing but the phantom of a dream. Still it is true that, beyond the limits of direct observation, our science is not infallible, and our theories and systems, although they may all contain a kernel of truth, undergo frequent changes, and are often revolutionized.

Through such a revolution the theory of chemistry has recently passed, and the system which is now universally accepted by the principal students of the science is greatly different from that which has been taught in our schools and colleges until within a few years. I have, therefore, felt that the best service I could render in this course of lectures would be to explain, as clearly as I am able, the principles on which the new philosophy is based, and to show in what it differs from the old. I have felt that there were many who, having studied what we must now call the old chemistry, would be glad to bridge over the gulf which separates it from the new, and to become acquainted with the methods by which we now seek to group together and explain the old facts.

Those who studied the science of chemistry twenty years ago, as it was taught, for example, in the works of the late Dr. Turner, were greatly impressed with the simplicity of the system and the beauty of its nomenclature. Until recently the study of the *new chemistry* has been far less inviting; since the science has been passing through a process of reconstruction, and displayed the imperfections of any half-built edifice; but it has now reached a condition in which it can be presented with the unity of a philosophical system. Our starting-point in the exposition of the modern chemistry must be the great generalization which is now known as the law of Avogadro, or Ampère. This law was first stated by Amedeo Avogádro, an Italian physicist, in 1811, and was reproduced by Ampère, a French physicist, in 1814. But, although attained thus early in the history of our science, this grand conception remained barren for nearly half a century. Now, however, it holds the same place in chemistry that the law of gravitation does in astronomy, though, unlike the latter, it was announced half a century before the science was sufficiently mature to accept it. The law of Avogadro may be enunciated thus:

EQUAL VOLUMES OF ALL SUBSTANCES, WHEN IN THE STATE OF GAS, AND UNDER LIKE CONDITIONS, CONTAIN THE SAME NUMBER OF MOLECULES (*Avogadro*, 1811—*Ampère*, 1814).

The enunciation of this law is very simple, but, before we can comprehend its meaning, we must understand what is meant by the term MOLECULE. This word is the one selected by Avogadro in the enunciation of his law. It is obviously of Latin origin, and means simply a *little mass* of matter. Ampère used in its place the word *particle*, in precisely the same sense. Both words signify the smallest mass into which any substance is capable of being subdivided by physical processes; that is, by processes which do not change its chemical nature. In many of our text-books it is defined as the smallest mass of any substance which can exist by itself, but both definitions are in essence the same.

As this is a very important point, it must be fully illustrated. In the first place, we recognize in Nature a great variety of different substances. Indeed, on this fact the whole science of chemistry rests; for, if Nature were made out of a single substance, there could be no chemistry, even if there could be intelligences to study science at all. Chemistry deals exclusively with the relations of *different* substances. Now, these substances present themselves to us under three conditions: those of the *solid*, the *liquid*, and the *gas*. Some substances are only known in one of these conditions, others in only two, while very many may be made to assume all three. Charcoal, for example, is only known in the solid state; alcohol has never been frozen, but can easily be volatilized; while, as every one knows, water can most readily be changed both into solid ice and into aëriform steam. Let me begin with this most familiar of all substances to illustrate what I mean by the word molecule.

When, by boiling under the atmospheric pressure, water changes into steam, it expands 1,800 times; or, in other words, one cubic inch of water yields one cubic foot of steam, nearly. Now, two suppositions are possible as modes of explaining this change.

The first is, that, in expanding, the material of the water becomes diffused throughout the cubic foot, so as to fill the space *completely* with the substance we

Fig. 1.

call water, the resulting mass of steam being absolutely homogeneous, so that there is no space within the cubic foot, however minute, which does not contain its proper proportion of water.

The second is, that the cubic inch of water consists of a certain number of definite particles, which, in the process of boiling, are not subdivided, so that the cubic foot of steam contains the same number of the same particles as the cubic inch of water, the conversion of the one into the other depending simply on the action of heat in separating these particles to a greater distance. Hence the steam is not absolutely homogeneous; for, if we consider spaces sufficiently minute, we can distinguish between such as contain a particle of water and those which lie between the particles. Now, the small masses of water, whose isolation we here assume, are what Avogadro calls molecules, and, follow-

Fig. 2.

ing his authority, we shall designate them hereafter exclusively by this word.

The rude diagrams before you will help me to make clear the difference between the two suppositions I have made. In the first (Fig. 1), we assume that the material of this cubic inch is uniformly expanded through the cubic foot. In the other (Fig. 2), we have in both volumes a definite number of molecules, the only difference being that these dots, which we have used to represent the molecules, are more widely separated in the one case than in the other. Now, which of these suppositions is the more probable? Let us submit the question to the test of experiment.

We have here a glass globe, provided with the necessary mountings—a stop-cock, a pressure-gauge, and a thermometer—and which we will assume has a capacity of one cubic foot. Into this globe we will first pour one cubic inch of water, and, in order to reduce the conditions to the simplest possible, we will connect the globe with our air-pump, and exhaust the air, although, as it will soon appear, this is not necessary for the success of our experiment. Exposing, next, the globe to the temperature of boiling water, all the liquid will evaporate, and we shall have our vessel filled with ordinary steam. If, now, that cubic foot of space is really packed close with the material we call water—if there is no break in the continuity of the aqueous mass—we should expect that the vapor would fill the space, to the exclusion of every thing else, or, at least, would fill it with a certain degree of energy which must be overcome before any other vapor could be forced in. Now, what is the case? The stop-cock of the globe is so arranged that we can introduce into it an additional quantity of any liquid on which we desire to experiment, without otherwise opening the vessel. If, then, by this means, we add more water, the additional quantity thus added will not evaporate, provided that the temperature remains at the boiling-point. Let us next, however, add a quantity of alcohol, and what do we find? Why, not only that this immediately evaporates, but we find that just as much alcohol-vapor will form as if no steam were present. The presence of the steam does not interfere in the least degree with the expansion of liquid alcohol into alcohol-vapor. The only difference which we observe is, that the alcohol expands more slowly into the aqueous vapor than it would into a vacuum. If, now that the globe is filled with aqueous vapor and alcohol-vapor at one and the same time, each acting, in all respects, as if it occupied the space alone, we add a quantity of ether, we shall have the same phenomena repeated. The ether will expand and fill the space with its vapor, and the globe will hold just as much ether-vapor as if neither of the other two were present; and so we might go on, as far as we know, indefinitely. There is not here a chemical union between the several vapors, and we cannot in any sense regard the space as filled with a compound of the three. It contains all three at the same time, each acting as if it were the sole occupant of the space; and that this is the real condition of things we have the most unquestionable evidence.

You know, for example, that a vapor or gas exerts a certain very considerable pressure against the walls of the containing vessel. Now, each of these vapors exerts its own pressure, and just the same pressure as if it occupied the space alone, so that the total pressure is exactly the sum of the three partial pressures.

Evidently, then, no vapor completely fills the space which it occupies, although equally distributed through it; and we can give no satisfactory explanation of the phenomena of evaporation except on the assumption that each substance is an aggregate of particles, or units, which, by the action of heat, become widely separated from each other, leaving very large intermolecular spaces, within which the particles of an almost indefinite number of other vapors may find place. Pass now to another class of facts, illustrating the same point.

The three liquids, water, alcohol, and ether, are expanded by heat like other forms of matter, but there is a striking circumstance connected with these phenomena, to which I wish to direct your observation. I have, therefore, filled three perfectly similar thermometer-bulb tubes, each with one of those liquids. The tubes are mounted in a glass cell standing before the condenser of a magic lantern, and you see their images projected on the screen. You also notice that the liquids (which have been colored to make them visible) all stand at the same height; and, since both the bulbs and the tubes are of the same dimensions, the relative change in volume of the inclosed liquids will be indicated by the rise or fall of the liquid columns in the tubes. We will now fill the cell with warm water, and notice that, as soon as the heat begins to penetrate the liquids, the three columns begin to rise, indicating an increase of volume; but notice how unequal is the expansion. The ether in the right-hand tube expands more than the alcohol in the centre, and that again far more than the water on the left. What is true of these three liquids is true in general of all liquids. Each has its own rate of expansion, and the amount in any case does not appear to depend on any peculiar physical state or condition of the liquid, but is connected with the nature of the substance, although, in what way, we are as yet wholly ignorant.

But you may ask: What is there remarkable in this? Why should we not expect that the rate of expansion would differ with different substances? Certainly, there is no reason to be surprised at such a fact. But, then, the remarkable circumstance connected with this class of phenomena has yet to be stated.

Raise the temperature of these liquids to a point a little above that of boiling water, and we shall convert all three substances into vapor. We thus obtain three gases, and, on heating these aëriform bodies to a still higher temperature, we shall find that, in this new condition, they expand far more rapidly than in the liquid state. But we shall also find that the influence of the nature of the substance on the phenomenon has wholly disappeared, and that, in the aëriform condition, these substances, and in general all substances, expand at the same rate under like conditions.

Why, now, this difference between the two states of matter? If the material fills space as completely in the aëriform as it does in the liquid condition, then we cannot conceive why the nature of the substance should not have the same influence on the phenomena of expansion in both cases. If, however, matter is an aggregate of definite small masses or molecules, which, while comparatively close together in the liquid state, become widely separated when the liquids are converted into vapor, then it is obvious that the action of the particles on each other, which might be considerable in the first state, would become less and less as the molecules were separated, until at last it was inappreciable; and if, further, as Avogadro's law assumes, the number of these particles in a given space is the same for all gases under the same conditions, then it is equally obvious that, there being no action between the particles, all vapors may be regarded as aggregates of the same number of isolated particles similarly placed, and we should expect that the action of heat on such similar masses would be the same.

Thus these phenomena of heat almost force upon us the conviction that the various forms of matter we see around us do not completely fill the spaces which they appear to occupy, but consist of isolated particles separated by comparatively wide intervals. There are many other facts which might be cited in support of the same conclusion; and among these two, which are more especially worthy of your attention, because they aid us in forming some conception of the size of the molecules themselves.

If this mass of glass is perfectly homogeneous—if the vitreous substance completely fills its allotted space, and there is no break whatever in the continuity of the material—then you would expect that its physical relations would not depend at all on the size of the surface affected. Suppose you wished to penetrate it with a fine wire. The point of this wire, however small, would not detect any difference at different points of the surface. Assume, however, that it consists of masses separated by spaces, like, for example, this sheet of wire netting. Then, although the surface would seem perfectly homogeneous to a bar large enough to cover a number of meshes, it would not be found to be by any means homogeneous to a wire which was small enough to penetrate the meshes. If, now, there are similar interstices in this mass of glass, we should expect that, if our wire were small enough (that is, of dimensions corresponding to the interstices), it would detect differences in the resistance at different points of this glass surface.

Make, now, a further supposition. Assume that we have a number of these wires of different sizes, the largest being twice as stout as the smallest. It is obvious that, if the interstices we have assumed were, say, several thousand times larger than the largest wire, all the wires would meet with essentially the same opposition when thrust at the glass. If, however, the interstices were only four or five times larger than the wires, then the larger would encounter much greater resistance from the edges of the meshes than the smaller.

It is unnecessary to say that no physical point can detect an inequality in the surface of a plate of glass, but we have, in what we call a beam of light, an agent which does find a passage through its mass. Now, it is perfectly true that we have no absolute knowledge of the nature of a beam of light. We have a very plausible theory that the phenomena of light are the effects of waves transmitted through a highly-elastic medium we call ether, and that, in the case of our plate of glass, the motion is transmitted through the ether, which fills the interstices between the molecules of this transparent solid; but we have no right to assume this theory in our present discussion.

Indeed, I cannot agree with those who regard the wave-theory of light as an established principle of science. That it is a theory of the very highest value I freely admit, and that it has been able to predict the phases of unknown phenomena, which experiment has subsequently brought to light, is a well-known fact. All this is true; but then, on the other side, the theory requires a combination of qualities in the ether of space, which I find it difficult to believe are actually realized. For instance, the rapidity with which wave-motion is transmitted depends, other things being equal, on the elasticity of the medium. Assuming that two media have the same density, their elasticities are proportional to the squares of the velocities with which a wave travels. The velocity of the sound-wave in air is about 1,100 feet a second or $\frac{1}{5}$ of a mile, that of the light-wave about 192,000 miles a second, or about one million times greater; and, if we take into account certain causes, which, though they tend to increase the velocity of sound, can have no effect on the luminiferous ether, the difference would be even greater than this.

Now, were the density of the ether as great as that of the atmosphere (say $\frac{1}{3}$ of a grain to the cubic inch), its elasticity or power of resisting pressure would be a million square, or a million million times that of the atmosphere. But, as you well know, the atmosphere can resist a pressure of about fifteen pounds to the square inch; hence the ether, when equally dense, would resist a pressure of fifteen million million pounds to the square inch, or, making the correction referred to above, seventeen million million pounds to the square inch. Of course, such numbers convey no impression, except that of vast magnitude; and you will obtain a clearer idea of the power when I tell you that this pressure is about the weight of a cubic mile of granite rock. Here is a glass cylinder filled with air, and here a piston which just fits it. The area of the piston is about a square inch—we will assume that it is exactly that. If we put a weight of fifteen pounds on the top of the piston, it will descend just half-way in the tube, and the air will be condensed to twice its normal density. Now, if we had a cylinder and piston, ether-tight as this is air-tight, and of sufficient strength, and, if we put on top of it a cubic mile of granite rock, it would only condense the ether to about the same density as that of the atmosphere at the surface of the earth. Of course, the supposition is an absurdity, for it is assumed that the ether pervades the densest solids as readily as water does a sponge, and could not, therefore, be confined; but the illustration will give you an idea of the nature of the medium which the undulatory theory assumes. It is a medium so thin that the earth, moving in its orbit 1,100 miles a minute, suffers no perceptible retardation, and yet endowed with an elasticity in proportion to its density a million million times greater than air.

Whether, however, there are such things as waves of ether or not, there is something concerned in the phenomena of light which has definite dimensions, that have been measured with as much accuracy as the dimensions of astronomy, although they are at the opposite extreme of the scale of magnitude. We represent these dimensions to our imagination as wave-lengths, that is, as the distances from crest to crest of our assumed ether-waves, and we shall find it difficult to think clearly upon the subject without the aid of this wave-theory, and every student of physics will bear me out in the statement that, though our theory may be a phantom of our scientific dreaming, these magnitudes must be the dimensions of something. Here they are:

Dimensions of Light-waves.

COLORS.	Number of waves in one inch.	Number of oscillations in one second.
Red	39,000	477,000,000,000,000
Orange	42,000	506,000,000,000,000
Yellow	44,000	535,000,000,000,000
Green	47,000	577,000,000,000,000
Blue	51,000	622,000,000,000,000
Indigo	54,000	658,000,000,000,000
Violet	57,000	699,000,000,000,000

You know that the sensation we call white light is a very complex phenomenon, and is produced by rays of all colors acting simultaneously on the eye. A very pretty experiment will illustrate this point. I have projected on the screen the image of a circular disk made of sectors of gelatine-paper, variously colored. By means of a very simple apparatus, I can revolve the disk, and thus cause the several colors to succeed each other at the same point with great rapidity, and you notice that the confused effect of the different colors produces the impression you call white, or, at least, nearly that.

The sunbeam produces the same impression, because it contains all these colored rays; and, if we pass it through a prism, the several rays, being bent unequally by the glass, diverge on emerging, so that, if we receive the beam thus divided on a screen placed at a sufficient distance, we obtain that magnificent band of blending hues we call the solar spectrum.

To each of the colored rays which fall along the line of the spectrum corresponds a definite wave-length. In the diagram, we have given the wave-lengths, corresponding to only a few selected points, one in each color, and marked in the solar spectrum itself by certain remarkable dark lines by which it is crossed. These values always create a smile with a popular audience, which makes it evident that, by those unfamiliar with the subject, they are looked upon as unreal if not absurd. But this is a prejudice. In our universe the very small is as real as the very great; and if science in astronomy can measure distances so great that this same swift messenger, light, traveling 192,000 miles a second, requires years to cross them, we need not be surprised that, at the other end of the scale, it can measure magnitudes like these.

Let not, then, these numbers impair your confidence in our results; but remember that the microscope reveals a universe with dimensions of the same order of magnitude. Moreover, the magnitudes with which we are here dealing are not beyond the limits of mechanical skill. It is possible to rule lines on a plate of glass so close together that the bands of fine lines thus obtained cannot be resolved even by the most powerful microscopes; and I am informed that the German optician, Nobert, has ruled bands containing about 224,000 lines to the inch. He regularly makes plates with bands consisting of from about 11,000 to 112,000 lines to the inch. These bands are numbered from the 1st to the 19th, and are used for microscopic tests. I am indebted to our friend Mr. Stodder for the opportunity of exhibiting to you a beautiful photograph of the 19th band, containing over 112,000 lines to the inch (Fig. 3). The photograph was made with one of Tolles's

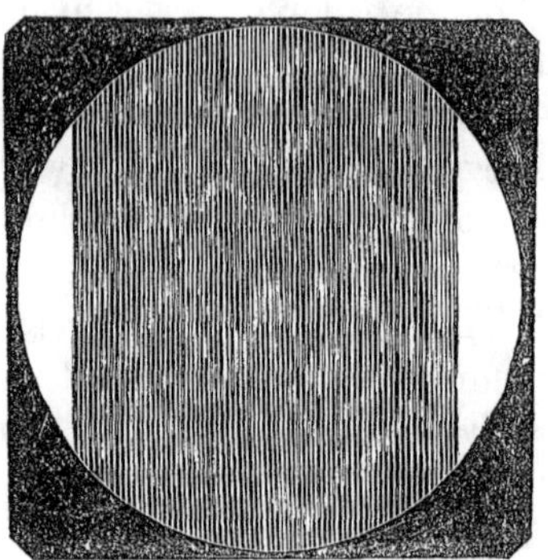

FIG. 3.—Nobert's 19th Band.

microscopes, and any microscopist will tell you that to resolve this band is a great triumph of art, and that you could have no better evidence of the skill of our eminent optician than this photograph affords. In projecting the image on the screen, some of the sharpness is lost, but I think the separate lines of the band must be distinctly visible to all who are not too far off.

Now, the distance between the lines on the original plate is not very different from one-half of the mean length of a wave of violet light, or one-third of a wave-length of red light; and, what is still more to the purpose, these very bands give us the means of measuring the dimensions of the waves of light themselves. Evidently, then, the dimensions with which we are dealing are not only conceivable, but wholly within the range of our perceptions, aided as they have been by the appliances of modern science.

But, to return to my argument: these values, if they are not wave-lengths, are real magnitudes, which differ from each other in size just as the above measurements show. Moreover, we have reason to believe that the various color-giving rays differ in nothing else, and it is certain from astronomical evidence that they all pass through the celestial spaces with the same velocity. Now, when a beam of light enters a mass of glass, not only does its velocity diminish, but, what is more remarkable, the different rays assume at once different velocities, and, according to the well-known principles of wave-motion, the unequal bending that results is the necessary effect of the unequal change in velocity which the rays experience. But, if the material of the glass were perfectly homogeneous throughout, it is impossible to conceive, either on the wave theory or any other theory of light we have been able to form, how a mere difference in size in what we now call the luminous waves should determine this unequal velocity with the accompanying difference of refrangibility, and the fact that such a difference is produced is thought by many to be strong evidence that there is not an absolute continuity in the material; in fine, that there are interstices in the glass, although they are so small that it requires the tenuity of a ray of light to detect them.

Still we cannot make our conceptions the measure of the resources of Nature, and I, therefore, do not attach much value to this additional evidence of the molecular structure of matter. But the importance of these optical phenomena lies in this, that, assuming the other evidence sufficient, they give us a rough measure of the size of the molecules. For, as is evident from our illustration with the wire meshes, the size of the molecular spaces cannot be very different from that of the waves of light. Our diagram shows that the red waves are only half as long again as the violet, and if the molecular spaces were, say, either ten thousand times larger or ten thousand times smaller than the mean length, the glass could produce no appreciable difference of effect on the different colored rays. We are thus led to the result that, if the glass is an aggregate of molecules, the magnitude of these molecules[1] is not very different from the mean length of a wave of light. Accepting the undulatory theory of light, we can submit the question, as Sir William Thompson has done, to mathematical calculation; and the result is that, though the effects of dispersion could not be produced unless the size of the molecules were far less than that of the wave-lengths, yet it is not probable that the size is less than say $\frac{1}{500,000,000}$ of an inch.

Before closing the lecture, allow me to dwell, for a few moments, on the second of the two classes of facts for which I have already bespoken your attention, since they confirm the results we have just reached, in a most remarkable manner. Every one has blown soap-bubbles, and is familiar with the gorgeous hues which they display. Many of you have doubtless heard that blowing soap-bubbles may be made more than a pleasant pastime, and I will endeavor to show how it can be made a philosophical experiment, capable of teaching some very wonderful truths. It is almost impossible to show the phenomena to which I refer to a large audience, and I cannot, therefore, feel any confidence in the success of the experiment which I am about to try; but I will show how you can all make the experi-

[1] The mean distance between the centres of contiguous molecules.

ment for yourselves. And, first, I must tell you how to prepare the soap-suds.

Procure a quart-bottle of clear glass and some of the best white castile-soap (or, still better, pure palm-oil soap). Cut the soap (about four ounces) into thin shavings, and, having put them into the bottle, fill this up with distilled or rain-water, and shake it well together. Repeat the shaking until you get a saturated solution of soap. If, on standing, the solution settles perfectly clear, you are prepared for the next step; if not, pour off the liquid and add more water to the same shavings, shaking as before. The second trial will hardly fail to give you a clear solution. Then add to two volumes of soap-solution one volume of pure, concentrated glycerine.

Those who are near can see what grand soap-bubbles we can blow with this preparation. The magnificent colors which are seen playing on this thin film of water are caused by what we call the interference of light. The color at any one point depends on the thickness of the film, and by varying the conditions we can show that this is the case, and make these effects of color more regular. For this purpose I will pour a little of the soap-solution into a shallow dish, and dip into it the open mouth of a common tumbler. By gently raising the tumbler it is easy to bring away a thin film of the liquid covering the mouth of the glass. You can all easily make the experiment, and study at your leisure the beautiful phenomena which this film presents. To exhibit them to a large audience is more difficult, but I hope to succeed by placing the tumbler before the lantern in such a position that the beam of light will be reflected by the film upon the screen, and then, on interposing a lens, we have at once a distinct image of the film. Success now depends on our keeping perfectly still, as the slightest jar would be sufficient to break this wonderfully delicate liquid membrane. See! the same brilliant hues which give to the soap-bubble its beauty are beginning to appear on our film, but notice that they appear in regular bands, crossing the film horizontally. As I have already stated, the color at any point depends on the thickness of the film, and, as it is here held in a vertical position, it is evident that the effect of gravity must be to stretch the liquid membrane, constantly thinning it out, beginning from the upper end—which, however, it must be remembered, appears on the screen at the lower end, since the lens inverts the image—and notice that, as the film becomes thinner and thinner, these bands of color which correspond to a definite thickness move downward, and are succeeded by others corresponding to a thinner condition of the film, which give place to still others in their turn. These colors are not pure colors, but the effect is produced by the overlapping of very many colored bands, and, in order to reduce the conditions to the simplest possible, we must use pure colored light—monochromatic light, as we call it. Such a light can be produced by placing a plate of red glass (colored by copper) in front of the lantern. At once all the particolors vanish and we have merely alternate red and dark bands. Watch, now, the bands as they chase each other, as it were, over the film, and notice that already new bands cease to appear, and that a uniform light tint has spread over the upper half (lower in the image) of the surface. Now comes the critical point of our experiment. If the film is in the right condition so that it can be stretched to a sufficient degree of tenuity, this light tint will be succeeded by a gray tint, ... and there it appears in irregular patches at the upper border. But in an instant all has vanished, for the film has broken, as it always breaks, soon after the gray tint appears.

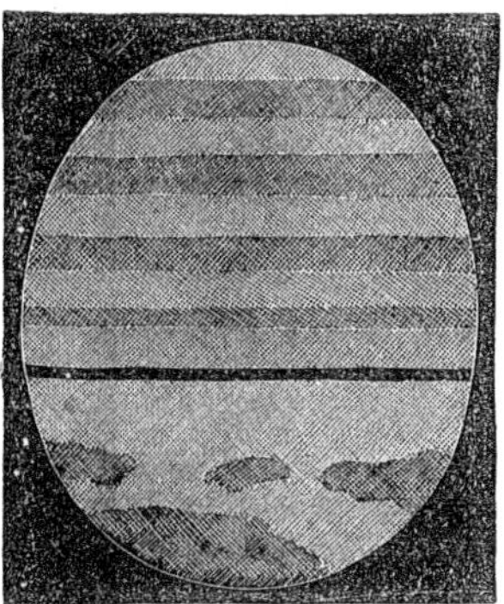

Fig. 4.—Bands on Soap-film.

Having now seen the phenomena, you will be better prepared to appreciate the strength of the argument to which I now have to ask your careful attention. You know that the red and dark bands seen in the last experiment, when we used the red glass, are caused by the interference of the rays of light, which are reflected from the opposite surfaces of the film. It is evident that the path of the rays reflected from the back surface must be longer than that of those reflected from the front surface by just twice the thickness of this film of water; and, as Prof. Tyndall has so beautifully shown you in the course of lectures just finished, whenever this difference of path brings the crests of the waves of one set of rays over the troughs of the second set, we obtain this wonderful result—that the union of the two beams of light produces darkness. It would, at first sight, seem that such a result must be produced in the case of our film whenever its thickness is equal to $\frac{1}{4}$, $\frac{3}{4}$, $\frac{5}{4}$, $\frac{7}{4}$, or any odd number of fourths of the length of a wave of red light, and this would be the case were it not for the circumstance that, in consequence of certain mechanical conditions, the rays of light reflected from the back of the film lose one-half of a wave-length in the very act of reflection. But, without entering into details, which have been so recently and so beautifully illustrated in this place, let me call your attention to this diagram, which tells the whole story:

ORDER OF BANDS.	Retardation of rays reflected from back-surface of film.	Thickness of film in waves of red light $\frac{1}{39000}$ of an inch.
Gray film...............	$\frac{1}{2}$ wave-length.	Less than $\frac{1}{4}$ wave-length.
Light film...............	1 "	$\frac{1}{4}$ "
First dark band........	$1\frac{1}{2}$ "	$\frac{1}{2}$ or $\frac{2}{4}$ "
First light band..........	2 "	$\frac{3}{4}$ "
Second dark band.......	$2\frac{1}{2}$ "	1 or $\frac{4}{4}$ "
Second light band.......	3 "	$1\frac{1}{4}$ " $\frac{5}{4}$ "
Third dark band.......	$3\frac{1}{2}$ "	$1\frac{1}{2}$ " $\frac{6}{4}$ "
Third light band........	4 "	$1\frac{3}{4}$ " $\frac{7}{4}$ "
Fourth dark band.......	$4\frac{1}{2}$ "	2 " $\frac{8}{4}$ "
Fourth light band.......	5 "	$2\frac{1}{4}$ " $\frac{9}{4}$ "

You thus see that the theory of light enables us to measure the thickness of the film, and we know that where that gray tint appeared in our experiment the thickness of the film was less than $\frac{1}{4}$ of the length of a wave of red light, or less than $\frac{1}{156,000}$ of an inch, and no wonder that the film broke when it reached such a degree of tenuity as that.

But, having followed me thus far, and being assured, as I hope you are, that we are on safe ground, and talking about what we do know, your curiosity will lead you to inquire whether we can stretch the film any farther.

The facts are that, after the appearance of the gray tint, although the film evidently stretches to a limited extent, it very soon breaks. Practically, then, we cannot stretch it beyond this point to any great extent; but why not? Theoretically, if the material of water is perfectly homogeneous, there would seem to be no good reason why it should not be capable of an indefinite extension, and why this film could not be stretched to an indefinite degree of attenuation. Assume, however, that water consists of molecules of a definite size, then it is evident that a limit would be reached as soon as the thickness of the film was reduced to the diameter of a single molecule. Obviously we could not stretch the film beyond this without increasing the distance between the molecules, and thus increasing the total volume of the water. Now, there is evidence that, when the gray tint appears, we are approaching a limit of this sort. It is hardly necessary to say that we cannot separate, to any considerable extent, the molecules of water from each other—that is, increase the distance between them—without changing the liquid into a gas, or, in other words, converting the water into steam, and the only way in which we can produce this effect is by the application of heat. The force required is enormous, but the force exerted by heat is adequate to the work, and it is one of the triumphs of our modern science that we have been able to measure this force, and reduce it to our mechanical standard. In order to pull apart the molecules of a pound of water, that is, convert it into steam, we must exert a mechanical power which is the equivalent of 822,600 foot-pounds, that is, a power which would raise nearly four tons to the height of one hundred feet, and, as we can readily estimate the weight of say one square-inch of our film, we know the force which would be required to pull apart the molecules of which it consists.

Again, on the other hand, singular as it may seem, we have been able to *calculate* the force which is required to stretch the film of water. This calculation is based on the theory of capillary action, of which the soap-bubble is an example. Moreover, to a certain limit, we are able to *measure* experimentally the force required to stretch the film, and we find that, as far as our experiments go, the theory and the experiments agree. Our experiments necessarily stop long before we reach the limit of the gray film; but our theory is not thus limited, and we can readily calculate how great a force would be required to stretch the film until the thickness was reduced to the $\frac{1}{500,000,000}$ of an inch; that is, the $\frac{1}{3,000}$ of the thickness of the light film, or the $\frac{1}{12,000}$ of a wave-length. Now, the force required to do this work is as great as that required to pull apart the molecules of the water and convert the liquid into vapor. It is therefore probable that, before such a degree of tenuity can be attained, a point would be reached where the film had the thickness of a single molecule, and that, in stretching it further, we should not reduce its thickness, but merely draw the molecules apart, and, thus overcoming the cohesion which determines its liquid condition, and gives strength to the film, convert the liquid into a gas.

There are many other physical phenomena which point to a similar limit, and, unless there is some fallacy in our reasoning, this limit would be reached at about the $\frac{1}{500,000,000}$ of an inch. Moreover, it is worthy of notice that all these phenomena point to very nearly the same limit. I have great pleasure in referring you, in this connection, to a very remarkable paper of Sir William Thompson, of Glasgow, on this subject, which, appearing first in the English scientific weekly called *Nature*, was reprinted in *Silliman's Journal* of July, 1870. He fixes the limits at between the $\frac{1}{250,000,000}$ and the $\frac{1}{5,000,000,000}$ of an inch, and, in order to give some conception of the degree of coarse-grainedness (as he calls it) thus indicated by the structure, he adds that, if we conceive a sphere of water as large as a pea to be magnified to the size of the earth, each molecule being magnified to the same extent, the magnified structure would be coarser-grained than a heap of small lead shot, but less coarse-grained than a heap of cricket-balls.

These considerations will, I hope, help to show you how definite the idea of the molecule has become in the mind of the physicist. It is no longer a metaphysical abstraction, but a reality, about which he reasons as confidently and as successfully as he does about the planets. He no longer connects with this term the ideas of infinite hardness, absolute rigidity, and other incredible assumptions, which have brought the idea of a limited divisibility into disrepute. His molecules are definite masses of matter, exceedingly small, but still not immeasurable, and they are the points of application to which he traces the action of the forces with which he has to deal. These molecules are to the physicist real magnitudes, which are no further removed from our ordinary experience on the one side, than are the magnitudes of astronomy on the other. In regard to their properties and relations, we have certain definite knowledge, and there we rest until more knowledge is reached. The old metaphysical question in regard to the infinite divisibility of matter, which was such a subject of controversy in the last century, has nothing to do with the present conception. Were we small enough to be able to grasp the molecules, we might be able to split them, and so, were we large enough, we might be able to crack the earth; but we have made sufficient advance since the days of the old controversy to know that questions of this sort, in the present state of knowledge, are both irrelevant and absurd. The molecules are to the physicist definite units, in the same sense that the planets are units to the astronomer. The geologist tears the earth to pieces, and so does the chemist deal with the molecules, but to the astronomer the earth is a unit, and so is the molecule to the physicist. The word molecule, which means simply a small mass of matter, expresses our modern conception far better than the old word atom, which is derived from the Greek *α*, privative, and *τέμνω*, and means, therefore, indivisible. In the paper just referred to, Sir W. Thompson used the word atom in the sense of molecule, and this must be borne in mind in reading his article. We shall give to the word atom an utterly different signification, which we must be careful not to confound with that of molecule. In our modern chemistry, the two terms stand for wholly different ideas, and, as we shall see, the atom is the unit of the chemist in the same sense that the molecule is the unit of the physicist. But we will not anticipate. It is sufficient for the present if we have gained a clear conception of what the word molecule means, and I have dwelt thus at length on the definition because I am anxious to give you the same clear conviction of their existence which I have myself. As I have said before, they are to me just as much real magnitudes as the planets, or, to use the words of Thompson, "pieces of matter of measurable dimensions, with shape, motion, and laws of action, intelligible subjects of scientific investigation."[1]

[1] *See* Lecture on Molecules, by Prof. Maxwell, *Nature*, Sept. 25, 1873.

LECTURE II.

THE MOLECULAR CONDITION OF THE THREE STATES OF MATTER—THE GAS, THE LIQUID, AND THE SOLID.

In my first lecture I endeavored to give you some conception of the meaning of the word molecule, and this meaning I illustrated by a number of phenomena, which not only indicate that molecules are real magnitudes, but which also give us some idea of their absolute size.

Avogadro's law declares that all gases contain, under like conditions of temperature and pressure, the same number of molecules in the same volume; and, if we can rely on the calculations of Thompson, which are based on the well-known theorem of molecular mechanics deduced by Clausius, this number is about one hundred thousand million million million, or 10^{23} to a cubic inch. Of course, as the volume of a given quantity of gas varies with its temperature and pressure, the number of molecules contained in a given volume must vary in the same way; and the above calculation is based on the assumption that the temperature is at the freezing-point, and the pressure of the air, as indicated by the barometer, thirty inches. The law only holds, moreover, when the substances are in the condition of perfect gases. It does not apply to solids or liquids, and not even to that half-way state between liquids and gases which Dr. Andrews has recently so admirably defined. In the state of perfect gas, it is assumed that the molecules are so widely separated that they exert no action upon each other, but the moment the gas is so far condensed that the molecules are brought within the sphere of their mutual attraction, then, although the aëriform state is still retained, we no longer find that the law rigidly holds; and when, by the condensation, the state of the substance is changed to that of a liquid or a solid, all traces of the law disappear. In order that you may gain a clear conception of this relation, I shall ask your attention in this lecture to the explanation which our molecular theory gives of the characteristic properties of the three conditions of matter, the gas, the liquid, and the solid. We begin with the gas, because its mechanical condition is, theoretically at least, by far the simplest of the three.

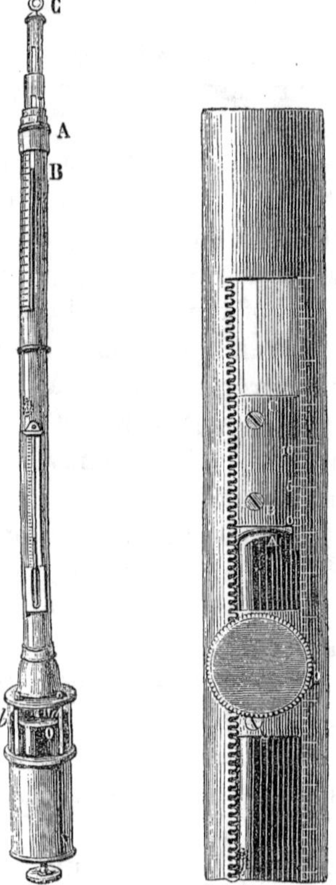

Fig. 5.—Barometer.

Every one of my audience must be familiar with the fact that every gas is in a state of constant tension, tending to expand indefinitely into space. In the case of our atmosphere, this tension is so great that the air at the level of the sea exerts a pressure of between fourteen and fifteen pounds on every square inch of surface—about a ton on a square foot.

It is this pressure which sustains the column of mercury in the tube of a barometer (Fig. 5); and since, by the laws of hydrostatics, the height of this column of mercury depends on the pressure of the air, rising and falling in the same proportion as the pressure increases or diminishes, we use the barometer as a measure of the pressure, and, instead of estimating its amount as so many pounds to the square inch, we more frequently describe it by the height in inches (or centimetres) of the mercury-column, which it is capable of sustaining in the tube of a barometer. The tension of the air is balanced by the force of gravitation, in consequence of which the lower stratum of the air in which we live is pressed upon by the whole weight of the superincumbent mass. The moment, however, the external pressure is relieved, the peculiar mechanical condition of the gas becomes evident.

Hanging under this large glass receiver is a small rubber bag (a common toy balloon), partially distended with air (Fig. 6). The air confined within the bag is exerting the great tension of which I have spoken, but the mass remains quiescent, because this tension is exactly balanced by the pressure of the atmosphere on the exterior surface of the bag. You see, however, that, as we remove, by means of this air-pump, the air from the receiver, and thus relieve the external pressure, the bag slowly expands, until it almost completely fills the bell. There can, then, be no doubt that there exists within this mass of gas a great amount of energy, and since this energy exactly balances the atmospheric pressure, it must be equal to that pressure.

But I wish to show you more than this, for not only

is it true that the bag expands as the pressure is relieved, but it is also true that the gas in the bag expands in exactly the same proportion as the external pressure

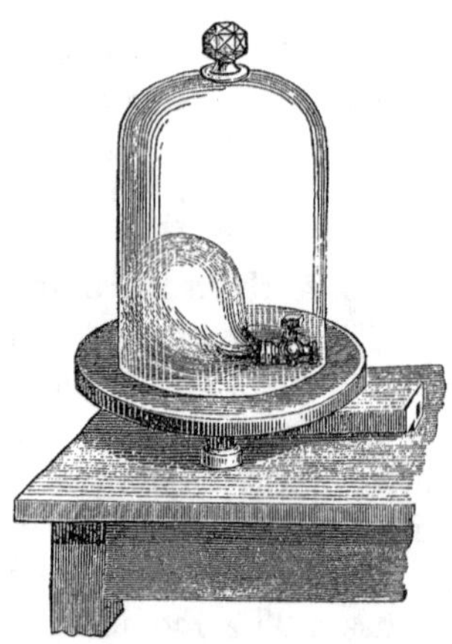

FIG. 6.—Expanding Bag under Air-pump.

diminishes. In order to prove this, I will now place under this same glass one of those small gasometers, which are used by the itinerant showmen in our streets for measuring what they call the volume of the lungs, while under this tall bell at the side I have arranged a barometer-tube for measuring the external pressure. The two receivers are connected together by rubber hose, so as to form essentially one vessel, and both are connected with the air-pump.

We will begin by blowing air into the gasometer until the scale marks 100 cubic inches, and, noticing after adjusting the apparatus that the barometer stands at 30 inches, we will now proceed to exhaust the air, at the same time carefully watching the barometer. . . . It has now fallen to 15 inches; that is, the pressure on the outside of the gasometer has been reduced to one-half, and the scale of the instrument shows me that the volume of the air in the interior has become 200 cubic inches; that is, has doubled. But let us continue the exhaustion. . . . The barometer now marks 10 inches, showing that the pressure has been reduced to one-third. The gasometer now contains 300 cubic inches of gas. The volume, then, has trebled. . . . Pushing the experiment still further, we have now the barometer standing at 7½ inches, and the scale of the gasometer shows that the volume of the inclosed air has become 400 cubic inches. The pressure has been reduced to one-fourth, and the volume of the air has quadrupled; and so we might go on. . . . Let, now, the atmosphere reënter the apparatus, and at once the air in the gasometer shrinks to its original volume, while the barometer goes back to 30 inches.

We might next take a condensing-pump, and, arranging our apparatus so as to resist the ever-increasing pressure, as the air was forced into the receivers, we should find that, when the barometer marked 60 inches, the scale of the gasometer would show 50 cubic inches, and that, when the mercury column had risen to 120 inches, the air in the gasometer would have shrunk to 25 cubic inches; and so on. There are, however, obvious mechanical difficulties, which make this phase of the experiment unsuitable for a large lecture-room, and what we have seen is sufficient to illustrate the general principle which I wished to enforce. The principle, in a few words, is this:

The volume of a confined mass of gas is inversely proportional to the pressure to which it is exposed: the smaller the pressure the larger the volume, and the greater the pressure the less the volume.

This principle holds true not only with air, but also with every kind of aëriform matter. If, instead of using that mixture of oxygen and nitrogen we call air, we had introduced into the gasometer 100 cubic inches of pure oxygen or of pure nitrogen, or of any other true gas, we should have obtained precisely the same effect. The results of the experiment are not in the least degree influenced by the nature of the gas employed; and, assuming that we start with the same gas-volumes, the resulting volumes are the same at each stage of the experiment. In every case the volume varies inversely as the pressure. The principle thus developed is one of the most important laws of physical science. It was discovered by the chemist Boyle in England in 1662, and verified by the Abbé Mariotte in France somewhat later, and is by some called the law of Mariotte, and by others the law of Boyle.

This law of Mariotte or Boyle is most closely related to the law of Avogadro. The one law is found to hold just as far as the other, and any deviation from the one is accompanied by a corresponding deviation from the other. So close, indeed, is the connection, that we cannot resist the conviction that the two laws are merely different phases of one and the same condition of matter; and our molecular theory explains this connection in the following way:

The molecules of a body are not isolated masses in a fixed position, all at rest, but, like the planets, they are in constant motion. In a gas this motion is supposed to take place in straight lines, the molecules hurrying to and fro across the containing vessel, striking against its walls, or else encountering their neighbors, rebounding and continuing on their course in a new direction, according to the well-known laws which govern the impact of elastic bodies. Of course, in such a system, all the molecules are not moving with the same velocity at the same time; but they have a certain mean velocity, which determines what we call the temperature of the body, and the higher the temperature the greater is this mean velocity; moreover, the mean velocity of the molecules of each substance is always the same at the same temperature. It varies, however, for different substances, and, for any given temperature, the less the density of the gas the greater is this velocity, although, as we shall hereafter see, the velocities of the molecules of two different gases are inversely proportional, not simply to their densities, but to the square roots of these quantities. We are able to calculate for each gas at least approximately what this velocity must be for any temperature, and, in the case of hydrogen gas, the value at the temperature of freezing water is about 6,097 feet per second. The internal energy, therefore, in a pound of hydrogen gas at the freezing-point is as great as that of a pound-ball moving 6,097 feet per second, and the energy in an equal volume (a little over 6.6 cubic yards when the barometer is at 30 inches) of any other true gas is equally great under the same conditions; a greater molecular weight compensating in every case for a less molecular velocity. Let us now bring together the two remarkable results already reached in this lecture.

One cubic inch of every gas, when the barometer marks 30 *inches, and the thermometer* 32° *Fahr., contains* 10^{23} *molecules.*

Mean velocity of hydrogen molecules, under same conditions, 6,097 *feet per second.*

It is evident, then, that every mass of gas must contain a large amount of internal energy, and this energy is made manifest in many ways, especially in what we call the permanent tension of the gas. Every surface in contact with a mass of gas is being constantly bombarded by the molecules, and hence the great pressure which results. Now, it can easily be seen that, if the volume of the gas is diminished—that is, if the same number of molecules are crowded into a less space—they will strike more frequently against a given surface, and therefore exert a greater pressure. Moreover, it can readily be proved, although the mathematical demonstration would be out of place in a popular lecture, that the pressure must be inversely as the volume; in other words, that the law of Mariotte is the necessary result of the molecular condition we have described.

Another effect of molecular motion is that condition of matter which the word temperature, just used, denotes. There are few scientific terms more difficult to define than this common word temperature. In ordinary language we apply the terms hot or cold to other bodies according as they are in a condition to impart heat to, or abstract it from, our own, and the various degrees of hot or cold are what we call, in general, temperature. Two bodies have the same temperature if, when placed together, neither of them gives or loses heat; and, when, under the same conditions, one body loses while the other gains heat, that body which gives out heat is said to have the higher temperature.

Increased temperature tested in this way is found to be accompanied by an increase of volume, and we employ this change of volume as the measure of temperature. This is the simple principle of a thermometer. The essential part of this instrument is a glass bulb, connected with a fine tube, and filled with mercury to a variable point in the stem. The least change in the volume of the mercury is indicated by the rise of the column in the tube. Primarily, the thermometer is a very delicate measure of the change of volume of the inclosed liquid; secondarily, it becomes a measure of temperature. You know how the thermometer is graduated. We plunge it into a mass of melting ice and mark the point to which the mercury falls, and then we immerse it in free steam, and mark the point to which the column rises. We now divide the distance between these fixed points into an arbitrary number of equal spaces, and continue the divisions of the same size above and below our two standard points. In our common Fahrenheit scale this distance is divided into 180 parts, the freezing-point is marked 32°, and the boiling, of course, 212°; the zero of this scale being placed at the thirty-second division below the freezing-point. In our laboratories we generally use a scale in which this distance is divided into 100 parts, and the freezing-point marked 0°, the divisions below freezing being distinguished with a *minus*-sign. All this, however, is purely arbitrary, and the instrument merely gives us the means of comparing temperatures. Here, for example, are two bodies. We apply the thermometer first to one and then to the other. It rises in each case to 50°. The only information we have obtained is, that both bodies are at the same temperature corresponding to a certain volume of the mercury in our thermometer, a temperature which we have agreed to call 50°; and we can predict that, if the two bodies are brought together, no heat will pass from one to the other. We now apply the thermometer to a third body, and it rises to 100°. We thus learn, further, that the third body is at a higher temperature than the other two, and in a condition to transfer to them a part of its heat. We cannot, however, say that its temperature is twice as high, or that it has any definite relation to that of the other two bodies.

There is, however, a theoretical way of measuring temperature, which appears to lead to something more than a mere arbitrary comparison. Let us assume that we have a cylindrical tube, closed below, but open above (Fig. 7). Let us further assume that the air

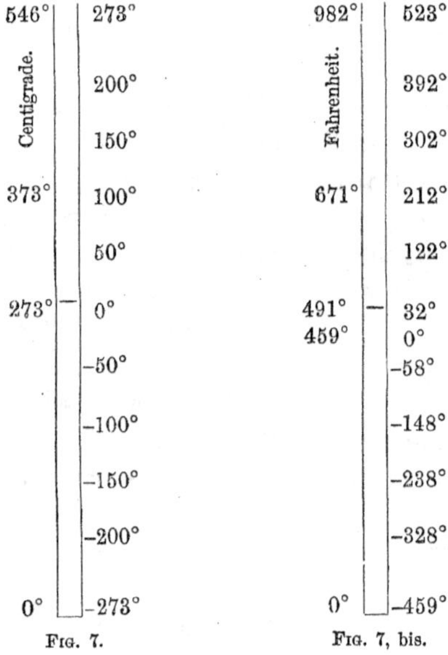

Fig. 7. Fig. 7, bis.

in the tube is confined by a piston, which has no weight, and moves without friction. As the temperature rises or falls, of course our assumed piston would rise or fall in the tube, following the expanding or contracting of the confined air. Let us mark the point to which the piston falls at the temperature of freezing water, 0°, and the point to which it rises at the temperature of boiling water, 100°. Lastly, let us divide the distance between these two points, as in a centigrade thermometer, into one hundred equal parts, and continue the divisions of the same size above 100° and below 0°. We shall find that we can make almost exactly 273 such divisions before reaching the closed bottom of our tube. Transfer, now, the zero of our scale to this lowest point or bottom of our tube, so that our old zero, or freezing-point of water, will be at 273° of the new scale, and the boiling-point of water at 373°.

We shall then have what is probably very nearly an absolute scale of temperature, such a one that we can say, for example, that the temperature at 500° is twice as great as that at 250°. Moreover, this is a scale such that the volume of any gas, under the same pressure, is exactly proportional to the temperature: for example, the volume of a given mass of air at 600° is twice as great as the volume at 300°. That this must be the case for air is evident from the construction of our theoretical thermometer; and it is equally true of any other perfect gas, for there would be no difference in effect whatever if the tube were filled with hydrogen, oxygen, or nitrogen, instead of air. It is very easy to refer degrees of our ordinary thermometer to degrees of this absolute scale. If the degrees are centigrade, we have merely to add 273; if they are Fahrenheit, we must add 459 (see Fig. 7, bis); and, for many purposes, it is exceedingly convenient to measure temperature in this way. Suppose, for example, we have 100 cubic inches of gas, at 4° centigrade, and we wish to know what would be its volume at 281°. Converting these values into absolute degrees by adding 273, we obtain 277° and 554°. Then, since the volume of a gas is exactly proportional to the absolute temperature, we have 277 : 554 = 100 : answer, 200 cubic inches. But the chief value of this method of measuring temperature is to be found in the simplicity with which it presents to us the property of gases we have been studying. The volume of a gas depends solely on two conditions: its pressure and its absolute temperature. As I before showed, it is inversely proportional to the pressure, and it now appears that it is directly proportional to the absolute temperature. We must then qualify the law of Mariotte by a second principle, equally fundamental and important:

The volume of a given mass of gas, under a constant pressure, varies directly as the absolute temperature.

This we call the law of Charles.

The molecular theory of gases explains the law of Charles very much in the same way as it explained the law of Mariotte. The pressure of a gas, as we have seen, is due to its molecular energy. If, by any means, we increase that energy, we must also increase the pressure in the same proportion; or, if the gas is free to expand under a constant pressure, we must increase the volume. In other words, the effect of increased energy must be the same as the effect which we know follows increased temperature. What more natural than to infer that the unknown condition, to which we have given the name of temperature, is simply molecular energy? Here, then, is our theoretical explanation of the law of Charles. The temperature of a body is the moving power of its molecules. At the 0° of our absolute scale the molecules would be reduced to a state of rest, and, at other temperatures, the molecular energy is directly proportional to the degrees of this scale; so that, for example, the molecules of air, at 273° (the 0° of centigrade), have only one-half of the energy which the same molecules possess when the temperature is raised to 546°. As the pressure exerted by the air must be proportional to the molecular energy, the increased temperature will, if the air is confined, double this pressure, or, if the air is free to expand under the constant pressure of the atmosphere, it will double the volume.

It would lead me too far to attempt to develop here at any greater length the dynamical theory of heat, and I regret that I am not able to do more than to give this bare outline of the remarkable properties of gases, which it so beautifully explains; but I take great pleasure in referring all who are interested in the subject to the very excellent work of Prof. Clerk Maxwell on the theory of heat. It is not a popular work, or one which is easy reading, but it contains a most elegant exposition of the modern theory of heat, in as simple a form as is consistent with accuracy and conciseness.

There is only one other point, in connection with the molecular theory of gases, to which it is important for me to refer in these lectures. We have seen that all gases have two essential characteristics: 1. Their volume is inversely proportional to the pressure to which they are exposed; and, 2. Their volume is directly proportional to the absolute temperature. Now, if we assume the molecular theory of gases as true, it can be proved, mathematically, that all gases at the same temperature and pressure must have the same number of molecules in the same volume. I do not give the proof, because it would be out of place here, and because all who are interested will find it in the work of

Prof. Maxwell, to which I have referred. It would be more satisfactory to enter into details, but I shall have accomplished the first object of this lecture if I have been able to leave with you a clear idea of the three laws which may be said to define the aëriform condition of matter, and which all true gases obey—

THE LAW OF MARIOTTE,
THE LAW OF CHARLES,
THE LAW OF AVOGADRO.

The first two are independent of any theory, and simply declare that the volume of every gas varies inversely as the pressure, and directly as the absolute temperature. The third is based on the molecular theory. It is more general, and includes the other two. It declares that equal volumes of all gases, under the same conditions of temperature and pressure, contain the same number of molecules.

Liquids are distinguished from gases chiefly in having a definite surface. Their particles have the same freedom of motion, but this motion is limited to the mass of the liquid. The particles of the air, if unconfined, would move off indefinitely into space; but the particles of this water, although moving with equal freedom within the liquid mass, cannot, as a rule, rise above what we call the surface of the water. Again, if we introduce a quantity of air, however small, into a vacuous vessel, it will instantly expand until it completely fills the vessel. A quantity of water, under the same conditions, will fall to the bottom of the vessel, and will be separated by a distinct surface from the vapor which forms above it. Lastly, if a gas is subjected to pressure, it is compressed in the exact proportion to the pressure, while with a liquid the compression is barely perceptible, even when the pressure is exceedingly great. Hence, gases are frequently called compressible and liquids incompressible fluids.

The explanation which the molecular theory gives of this difference of relations is very simple. In the gas the molecules are separated beyond the sphere of each other's influence, and move through space wholly free from the effects of the mutual attraction. In a liquid, on the other hand, this attraction, which we call cohesion, is very sensible, and restrains the individual molecules within the mass, although they are free to move among themselves. You can easily understand, by referring again to the diagram (Fig. 2, on page 16), how this attractive force would act.

A molecule, in the midst of the mass, moves freely, because the attractions are equal in all directions, but a molecule near the surface is in a very different condition. As it approaches the surface, the attraction toward the mass of the liquid becomes greater than the attraction toward the surface, and when it reaches the surface the whole force of the inward attraction is pulling it back, and, unless the moving power of the molecule is sufficiently great to overcome this force, its motion is arrested, and it turns back on its course. It may happen, however, especially when heat is entering the liquid, that some of the molecules, through the effects of their mutual collisions, acquire sufficient energy to fly off from the liquid mass, and hence result the well-known phenomena of evaporation. Thus our theory defines the liquid condition of matter, and explains how the liquid is converted by heat into the gas.

In all theoretical discussions, it is always highly satisfactory when, in following out our theoretical conceptions to their consequences, we find that these consequences are actually realized in natural phenomena, and such satisfaction we can have in the present case. Consider what must be the form which a mass of liquid molecules isolated in space would necessarily take. Remember that these molecules are moving with perfect freedom within the body, but that the extent of the motion of each molecule is limited by the attraction of the mass of the liquid. Remember also that, according to the well-known principles of mechanics, this attraction may be regarded as proceeding from a single point, called the centre of gravity. Remember, further, that the molecules have all the same moving power, and you will see that the extreme limits of their excursions to and fro through the liquid mass must be on all sides at the same distance from the central point. Hence the bounding surface will be that whose points are all equally distant from the centre. I need not tell you that such a surface is a sphere, nor that a mass of liquid in space always assumes a spherical form. The rain-drops have taught every one this truth. Still, a less familiar illustration may help to enforce it. I have therefore prepared a mixture of alcohol-and-water, of the same specific gravity as olive-oil, and in it I have suspended a few drops of the oil. By placing the liquid in a cell, between parallel plates of glass, I can readily project an image of the drops on the screen, and I wish you to notice how perfectly spherical they are. And I would have you, moreover, by the aid of your imagination, look within this external form, and picture to yourselves the molecules of oil moving to and fro through the drops, but always slackening their motion where they approach the surface, and on every side coming to rest and turning back at the same distance from the centre of motion.

Neither liquids nor gases present the least trace of structure. They cannot even support their own weight, much less sustain any longitudinal or shearing stress. A solid, on the other hand, has both tenacity and structure, and resists, with greater or less energy, any force tending to alter its form, as well as change its volume. The tenacity and peculiar forms of elasticity which solids exhibit are characteristics which are familiar to every one, but the evidences of structure are not so conspicuous. The structure of solids is most frequently manifested by their crystalline form, and this form is one of the most marked features of the solid state. But although, under definite conditions, most substances assume a fixed geometrical form, yet, to ordinary experience, these forms are the exceptions, and not the rule. I will therefore make the crystallization of solid bodies the subject of a few experimental illustrations.

For the first experiment, I have prepared a concentrated solution of ammonic chloride (sal-ammoniac), and with this I will now smear the surface of a small glass plate. Placing this before our lantern, and using a lens of short focus, so as to form a greatly-enlarged image on the screen, let us watch the separation of the solid salt as the solution evaporates. . . . Notice that, first, small particles appear, and then from these nuclei the crystals shoot out and ramify in all directions, soon covering the plate with a beautiful net-work of the filaments of the salt. We cannot here, it is true, distinguish any definite geometrical form; but it can be shown that these very filaments are aggregates of such forms, and their structure is made evident by a fact, to which I would especially call your attention—that, as the crystalline shoots ramify over the plate, the sprays keep always at right angles to the stem, or else branch at an angle of 45°, which is the half of a right angle (Fig. 8).

For a further illustration of the process of crystallization I have prepared a solution in alcohol of a solid

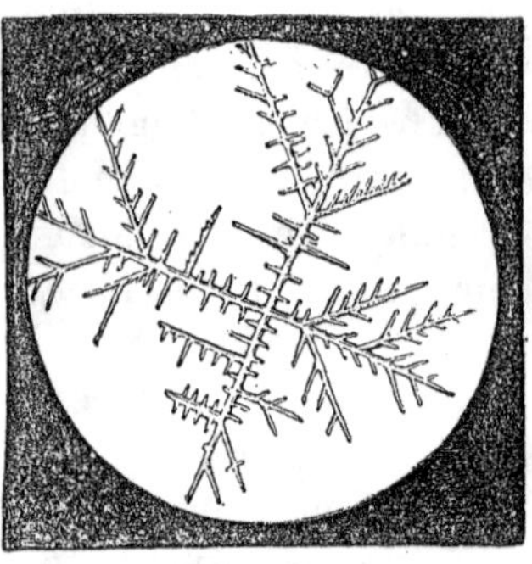

Fig. 8.—Crystallization of Sal-Ammoniac.

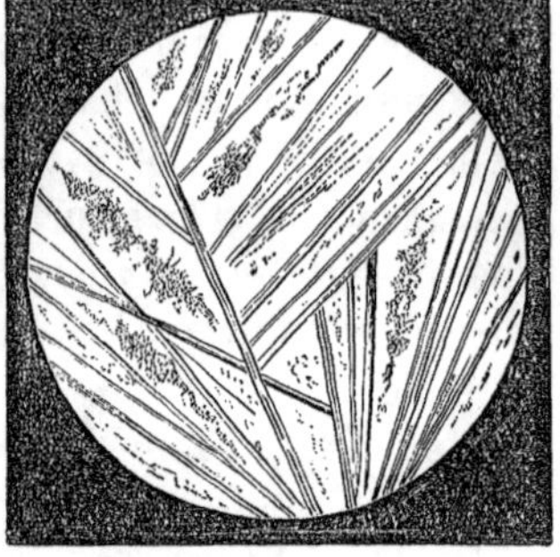

Fig. 9.—Crystallization of Urea.

substance called urea, with which we will experiment in precisely the same way as before. . . . The process of crystallization, which is here so beautifully exhibited, is one of the most striking phenomena in the whole range of experimental science. It is, of course, not so wonderful as the development of a plant or an animal from its germ, but then organic growth is slow and gradual, while here beautiful, symmetrical forms shape themselves in an instant out of this liquid mass, revealing to us an architectural power in what we call lifeless matter, whose existence and controlling influence but few of us have probably realized. The general order of the phenomena in this experiment is the same as in the last; but notice how different the details. We do not see here that tendency to ramify at a definite angle, but the crystals shoot out in straight lines, and cover the plate with bundles of crystalline fibres, which meet or intersect each other irregularly as the accidental directions of the several shoots may determine (Fig. 9). As before, we cannot recognize the separate crystals; indeed, large isolated crystals, such as you may see in collections of minerals, cannot be formed thus rapidly. They are of slow growth, and only found where the conditions have favored their development. But all the mineral substances, of which the rocks of our globe consist, have a crystalline structure, and are aggregates of minute crystals like the arborescent forms whose growth you have witnessed.

The external form is but one of the indications of crystalline structure, and by various means this structure may frequently be made manifest when the body appears wholly amorphous. Nothing could appear externally more devoid of structure than a block of transparent ice. Yet it has a most beautiful symmetrical structure, which can easily be made evident by a very simple experiment, originally devised, I believe, by Prof. Tyndall. For this purpose I have prepared a plate of ice about an inch in thickness, whose polished surfaces are parallel to the original plane of freezing. I will now place this plate in front of the condenser of my lantern, and, placing before it a lens, we will form on the curtain an image of the ice-plate, some twenty times as large as the plate itself. The rays of heat which accompany the light-rays of our lantern soon begin to melt the ice; but, in melting it, they also dissect it, and reveal its structure. . . . Notice those symmetrical six-pointed stars which are appearing on the wall (Fig. 10). Prof. Tyndall calls them, very appropriately, ice-flowers, for, as the flower shows forth the structure of the plant, so these hexagonal forms disclose the six-sided structure of ice. You can hardly fail to notice the similarity of these forms to those of the snow-flake. The six petals of the ice-flowers on our screen make with each other an angle of 60°, and, if you examine, with a magnifier, flakes of fresh-fallen snow (Fig. 11), or the arborescent

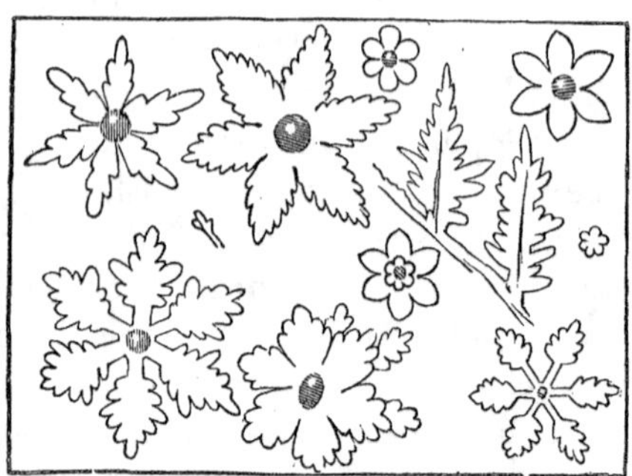

Fig. 10.—Ice-Flowers.

forms which crystallize on the window-panes in frosty weather, you will find that, in all cases, the crystalline shoots ramify at this angle, which is as constant a character of the solid condition of water as is the right angle of sal-ammoniac.

There are other solids whose crystalline structure, like that of ice, becomes evident during melting; but a far more efficient means of discovering the structure of solids, when transparent, is furnished by polarized light.

It would be impossible for me, without devoting a great deal of time to the subject, either to explain the nature of what the physicists call polarized light, or to give any clear idea of the manner in which it brings out the structure of the solid. I can only show you a few experiments, which will make evident that such is the fact. We have now thrown on the screen a luminous disk, which is illuminated by polarized light. To the unaided eye it does not appear differently from ordinary light; but there is this peculiarity in the beam. I have here a prism of well-known construction, made of Iceland-spar, and called a Nicol prism.

FIG. 11.—Snow-Crystals.

The spar is as translucid as glass, and, with ordinary light, it transmits, as you see, the beam equally well, whether it is placed in one position or another. But, with the polarized beam, we shall have a very different result. In one position, as you notice, it allows the light to pass freely; but, on turning it round through an angle of 90°, almost all the light is intercepted: the beam of light seems to have sides, which stand in a different relation to the prism in one position from that which they bear to it in the other. To describe this condition of the beam, the early experimenters adopted the word *polarized*, which was not, however, a happy designation; for the term now implies an opposition of relations very unlike the difference which we recognize between the sides of such a beam of light. Placing now the Nicol prism in the position in which it intercepts the polarized beam, I will first place between it and the source of light a plate of glass. You notice that there is no difference of effect. Besides the arrangement for polarizing the light and the Nicol prism there is no other apparatus here except a lens, which would form on the screen an image of the glass plate or of any thing depicted upon it, were it not for the circumstance that the Nicol prism cuts off the light. By turning the Nicol so that the polarized light can pass, and putting a glass photograph in the place of the glass plate, you see at once the photograph projected on the screen. Having turned back the Nicol until the light is again intercepted, I will remove the photograph, and put in its place a thin sheet of gypsum. . . . See this brilliant display of colors. The plate of gypsum is as colorless and transparent as the glass, and the gorgeous hues result from the decomposition of the polarized light produced by the crystalline structure of the gypsum. I will next turn round the film of gypsum, and you notice that the colors gradually fade out and finally disappear. As we turn farther they reappear, and so on. Evidently, the colors are only produced in a definite position of the gypsum plate with reference to our polarizing apparatus. Moreover, as I can readily show you, the tint of color depends on the thickness of the film. I have here a simple geometrical design formed of plates of gypsum of different thicknesses, and you notice that each plate assumes a different hue. On turning, however, our Nicol prism 90°, these colors are suddenly exchanged for their complementary tints.

It is obvious that any colored designs might be reproduced in this way by combining gypsum plates cut to the required thickness and form, as in mosaic work; and I will now show you a number of beautiful illustrations of this peculiar form of art. . . . But you cannot appreciate the wonder of these experiments without bearing in mind that these gypsum mosaics show no color whatever in ordinary light, consisting, as they do, of plates which appear like colorless glass.

Let me now substitute for the gypsum designs the glass plate on which we recently crystallized urea, and notice that the crystals of this substance, which we saw form on the glass, yield similar brilliant hues. The experiment becomes still more striking, if we crystallize the salt under these conditions. I will, therefore, take another glass plate, and, having smeared it as before with the solution of urea, I will place it in the focus of my lens before the polarizer. The field is now perfectly dark, but, as soon as the crystals begin to form, you see these colored needles shoot out on the dark ground, presenting a phenomenon of wonderful beauty.

Now, all this indicates a definite structure, and, to those familiar with these phenomena, they point to a definite conclusion in regard to this structure. I wish I could fully develop the argument before you, but this would require more time than the plan of my lectures allows, and I must be content if I have been able to impress upon your minds the single general truth which these experiments suggest. You saw the urea crystallize, that is, assume a definite structure, and you now see that this structure so modifies the polarized light as to produce these gorgeous hues. You have seen similar hues, but still more brilliant, produced by a plate of gypsum, and I can only add that the conclusion which the analogy suggests is legitimate, and sustained by the most conclusive evidence. The transparent plates of gypsum have as definite a structure as the crystals of urea, and to the student of optics these colors reveal that structure just as clearly as it is manifested, even to the uninstructed eye, by the processes of crystallization, which we have witnessed this evening.

Would, however, that I could convey to you a more

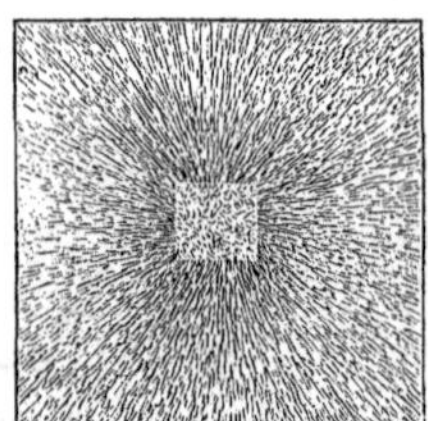

FIG. 12.—Magnetic Curves, one pole.

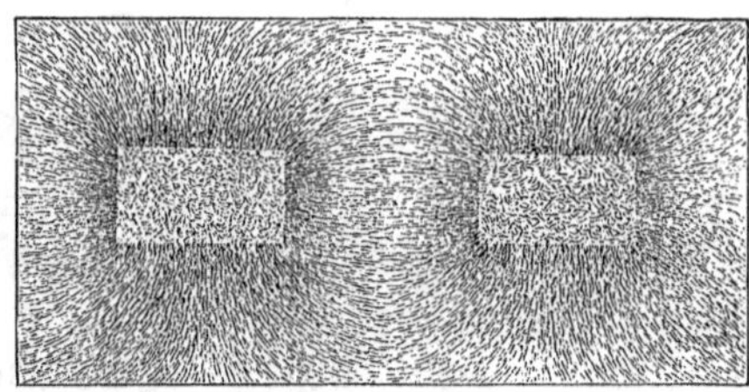

FIG. 13.—Magnetic Curves, two poles.

definite idea of the nature of that structure, for our theory gives us a very clear conception of what we suppose to be the relations of the molecules in these solid bodies! But the subject is a difficult one, and it would require a long time to make the matter intelligible. Still, by the aid of a few parallel experiments, I may be able to give you, at least, a glimpse of the manner in which, as we suppose, the structure of solid bodies is produced.

Everybody knows that a magnetic needle, when free to move, assumes a definite position, pointing, in general, north and south. Now, a magnetic needle is a needle of steel (hardened iron) in a condition which we call polarized; and, what is true of it, is true of every polarized body, to a greater or less extent. So, also, if we have a collection of such polarized bodies, they will always arrange themselves in some definite position with reference to each other—will form, in a word, a definite structure.

Further, it is well known that a magnet polarizes all masses of iron in its neighborhood, and this circumstance enables us to illustrate the truth of the principle just stated, in a most striking manner: If we bring a bar-magnet near some iron filings sprinkled over a

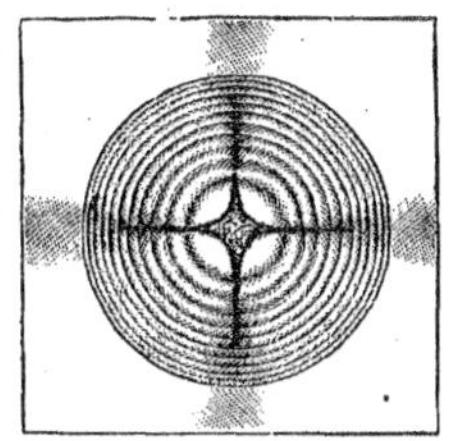

Fig. 14.—Rings, Uniaxial Crystals.

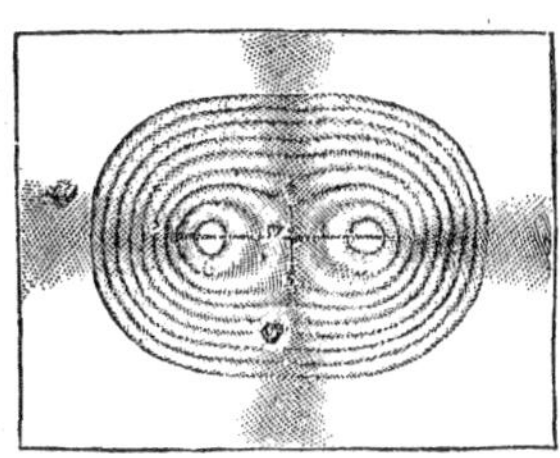

Fig. 15.—Rings, Biaxial Crystals.

plate of glass, these little bits of iron become at once polarized by induction; and, if then we gently tap the glass, the iron particles will swing round on its smooth surface, and arrange themselves in the most wonderful way. By means of my vertical lantern I can show you this effect most beautifully. I first sprinkle the filings on the glass stage of our lantern, and then, having protected them by a thin covering-glass, I bring near the glass one of the poles of a bar-magnet. . . . Notice how, on tapping the glass, the filings spring into position, arranging themselves on lines radiating from this pole (Fig. 12). Here, evidently, we have a definite structure produced. Let us now clear our stage, and arrange for a second experiment. This time, however, we will lay the bar-magnet on the covering-glass, so that the bits of iron shall be brought under the influence of both of its poles at the same time. . . . See what a beautiful set of curves results on tapping the glass (Fig. 13), and let me beg you to try to carry in your mind for a moment the general aspect of this structure, as well as of the first.

Now, we suppose that, in solid bodies, the structure depends on the polarity of the molecules, and that the molecules, like the bits of iron in our experiment, take up the relative position which the polar forces require. And, next, I will show you that a beam of polarized light develops in some solids an evidence of structure not very unlike that you have just seen.

Returning, then, to our polariscope, I place in the beam of light a plate of Iceland-spar cut in a definite manner. . . . See those radiating lines, and those iris-colored circles (Fig. 14). Does not that remind you of the structure we developed around a single magnetic pole? Next, I will use a similar plate cut from a crystal of nitre; . . . . and, see, we have almost the reproduction of the curves about the double pole (Fig. 15). It is the form of the curves as indicating a certain structure, not the brilliant colors, to which I would direct your attention. The iris hues are caused simply by the breaking up of the white light we are using; for the crystal decomposes it to a greater or less extent, like a prism. If, by interposing a plate of red glass, we cut off all the rays except those of this one color, the varied tints disappear, but, in the black curves which now take their place, the analogy I am endeavoring to present becomes still more marked. Certainly, you could have no more striking analogy than this. I can add nothing by way of commentary to the experiments without entering into unsuitable details, and I will only say, further, that I am persuaded that the resemblances we have seen have a profound significance, and that the structure, which the polarized beam reveals in these solid bodies, is really analogous to that which the magnet produces from the iron filings.

LECTURE III.

HOW MOLECULES ARE WEIGHED.

In order that we may make sure of the ground we have thus far explored, let me recapitulate the characteristic qualities of the three conditions of matter which I sought to illustrate in the last lecture.

A gas always completely fills the vessel by which it is inclosed. It is in a state of permanent tension, and conforms to the three laws of Mariotte, of Charles, and of Avogadro. A liquid has a definite surface. It can be only very slightly compressed, and obeys neither of these three laws. A solid has a definite structure, and resists both longitudinal and shearing stresses to a limited extent.

Having now presented to you the molecular theory as fully as I can without entering into mathematical details, I come back again to the great law of Avogadro, which is at the foundation of our modern chemistry:

When in the condition of a perfect gas, all substances, under like conditions of temperature and pressure, contain in equal volumes the same number of molecules.

I have already shown you that, if we assume the general truth of the molecular theory (in other words, if we assume that a mass of gas is an aggregate of isolated moving molecules), then the law of Avogadro follows as a necessary consequence from the known properties of aëriform matter, and may, therefore, in a certain limited sense, be said to be capable of proof. As yet, however, we have only considered the purely physical evidence in favor of the law. We come next to the chemical evidence which may be adduced in support of its validity, and this is equally strong.

It would be impossible at the present stage of our study to make the force of this evidence apparent, because, so far as chemistry is concerned, the law of Avogadro is a generalization from a large mass of facts, and the proof of its validity is to be found solely in the circumstance that it not only explains the known facts of chemistry, but that it is constantly leading to new discoveries. This law, as I have intimated, bears about the same relation to modern chemistry that the law of gravitation does to modern astronomy. Modern astronomy itself is the proof of the law of gravitation; modern optics the proof of the undulatory theory of light; and so the whole of modern chemistry, and nothing less, is the proof of the law of Avogadro. I do not say that this great law of chemistry stands as yet on as firm a basis as the law of gravitation; but I do say that it is based on as strong foundations as the undulatory theory of light, and is more fully established to-day than was the law of gravitation more than a century after it was announced by Newton. I have already briefly referred to the history of the law.

The original memoir was published by Amedeo Avogadro in the *Journal de Physique*, July, 1811. In this paper the Italian physicist "enunciated the opinion that gases are formed of material particles, sufficiently removed from one another to be free from all reciprocal attraction, and subject only to the repulsive action of heat;" and, from the facts, then already well established, that the same variations of temperature and pressure produce in all gases nearly the same changes of volume, he deduced the conclusion that equal volumes of all gases, compound as well as simple, contain, under like conditions, the same number of these molecules.

This conception, simple and exact as it now appears, was at the time a mere hypothesis, and was not advanced even with the semblance of proof. The discovery of Gay-Lussac, that gases combine in very simple proportions by volume, was made shortly after, and, had its important bearings been recognized at once, it would have been seen to be a most remarkable confirmation of Avogadro's doctrine. But the new ideas passed almost unnoticed, and were reproduced by Ampère in 1814, who based his theory on the experiments of Gay-Lussac, and defended it with far weightier evidence than his predecessor. Still, even after it was thus reaffirmed, the theory seems to have received but little attention either from the physicists or the chemists of the period. The reason appears to have been that the *integrant molecules* of Avogadro and the *particles* of Ampère were confused with the *atoms* of Dalton, and, in the sense which the chemists of the old school attached to the word atom, the proposition appeared to be true for only a very limited number even of the comparatively few aëriform substances which were then known. Moreover, the atomic theory itself was rejected by almost all the German chemists; and, in physics, the theory of a material caloric then prevailing was not enforced by the new doctrine. In a word, this beautiful conception of Avogadro and Ampère came before science was ripe enough to benefit by it. A half-century, however, has produced an immense change. The development of the modern theory of chemistry has made clear the distinction between molecules and atoms, while the number of substances known in their aëriform condition has been vastly increased. It now appears that, with a few exceptions, all these substances conform to the law, and these exceptions can, for the most part at least, be satisfactorily explained. On the other side, in the science of physics, more exact notions of the principles of dynamics have become general, and the dynamical theory of heat necessarily involves the law of equal molecular volumes. Thus, this theory of Avogadro and Ampère, which remained for half a century almost barren, has come to stand at the diverging-point of two great sciences, and is sustained by the concurrent testimony of both. It is not, then, without reason that we take this law as the basis of the modern system of chemistry; and, starting from it, let us see to what it leads:

In the first place, then, it gives us the means of determining directly the relative weight of the molecules of all such substances as are capable of existing in the aëriform condition. For, it is obvious, *if equal volumes of two gases contain the same number of molecules, the relative weights of these molecules must be the same as the relative weights of the equal gas-volumes.* Thus, a cubic foot of oxygen weighs sixteen times as much as a cubic foot of hydrogen under the same conditions. If, then, there are in the cubic foot of each gas the same number of molecules, each molecule of oxygen must weigh sixteen times as much as each molecule of hydrogen.

It is much more convenient in all chemical calculations to use the French system of weights and measures; and since, through modern school-books, the names of these measures have become quite familiar to almost every one, I think I can refer to them without confusion. The accompanying table will serve to refresh your memory, and may be useful for reference:

The metre is approximately the $\frac{1}{10,000,000}$ *part of a quadrant of a meridian of the earth measured from the pole to the equator.*

The metre equals 10 *decimetres or* 100 *centimetres.*

The cubic metre, or stère, equals 1,000 *cubic decimetres or litres.*

The cubic decimetre, or litre, equals 1,000 *cubic centimetres.*

The gramme is the weight, in vacuo, of one cubic centimetre of water at 4° *centigrade (the point of maximum density).*

The kilogramme equals 1,000 *grammes, and is, therefore, the weight of one cubic decimetre or litre of water under the same conditions.*

The crith is the weight, in vacuo, of one litre of hydrogen gas at 0° *centigrade (the freezing-point of water), and at* 76 *centimetres (the normal height of the barometer). It equals* 0.09 *of a gramme very nearly.*

The metre is equal to $3\frac{1}{4}$ *feet nearly.*

The litre is equal to $1\frac{3}{4}$ *pint nearly.*

The gramme is equal to $15\frac{1}{2}$ *grains nearly.*

The kilogramme is equal to $2\frac{1}{5}$ *pounds nearly.*

The convenience of the French system depends not at all on any peculiar virtue in the metre (the standard of length on which the system is based), but upon the two circumstances—1. That all the standards are divided decimally so as to harmonize with our decimal arithmetic; and, 2. That the measures of length, volume, and weight, are connected by such simple relations that any one can be most readily reduced to either of the other two. In order to make clear these last relations, I must ask you to distinguish between two terms which are constantly confounded in the ordinary use of language, namely, density and specific gravity.

The density of a substance is the amount of matter in a unit-volume of the substance. In the English system it is the weight in grains of a cubic inch, and in the French system the weight in grammes of a cubic centimetre. Thus the density of wrought-iron is 1,966 grains English, or 7.788 grammes French. So also the density of water at 4° centigrade (the point of maximum density) is 252.5 grains, or 1 gramme.

The specific gravity of a substance is the ratio between the weight of the substance and that of an equal volume of some other substance taken as a standard. For liquids and solids, water is always the standard selected, and the specific gravity, therefore, expresses how many times heavier the substance is than water. It can evidently be found by dividing the density of the substance by the density of water, because, as we have just seen, these densities are the weights of equal volumes. Hence the specific gravity of iron equals—

$$\frac{1966 \text{ grains}}{252.5 \text{ grains}} \text{ or } \frac{7.788 \text{ grammes}}{1 \text{ gramme}} = 7.788$$

Of course, the specific gravity of a substance will be expressed by the same number in all systems; and, further, in the French system, as the example just cited shows, this number expresses the density as well as the specific gravity. Density, however, is a weight, while specific gravity is a ratio, and the two sets of numbers are identical in the French system only because in that system the cubic centimetre of water has been selected as the unit of weight.

In the French system, then, the same number expresses both the specific gravity and also the weight of one cubic centimetre of the substance in grammes; and, since both 1,000 grammes = 1 kilogramme, and 1,000 cubic centimetres = 1 litre, it expresses also the weight of one litre in kilogrammes. These relations are shown in the following table:

The specific gravity of a liquid or solid shows how many times heavier the body is than an equal volume of water at 4° centigrade. The same number expresses also the weight of one cubic centimetre of the substance in grammes, or of one litre in kilogrammes.

	Alcohol.	Water.	Sulphur.	Iron.	Gold.
Sp. Gr.,	0.8	1.	2.1	7.8	19.3
Density,	0.8 gram.	1. gram.	2.1 gram.	7.8 gram.	19.3 gram.

The black squares are supposed to represent cubic centimetres. If assumed to represent cubic decimetres, then the weights which measure the densities would be in kilogrammes instead of grammes. It will now be seen how simple it is in the French system to calculate weight from volume. When the specific gravity of a substance is given, we know the weight both of one cubic centimetre and of one litre of that substance, and we have only to multiply this weight by the number of cubic centimetres, or of litres, to find the weight of the given volume. Thus the weight of a wrought-iron boiler-plate ½ centimetre thick, and measuring 120 centimetres by 75, would be—

$$0.5 \times 120 \times 75 \times 7.788 = 35{,}046 \text{ grammes.}$$

In general—

$$W. = V. \times \text{Sp. Gr.}$$

When V. is given in cubic centimetres, the resulting weight will be in grammes; when in litres, the weight will be kilogrammes.

In estimating the specific gravity of gases, we avoid large and fractional numbers, by selecting, as our standard, hydrogen gas, which is the lightest form of matter known; but we thus lose the advantage gained by having the unit-volume of our standard the unit of weight. It is no longer true that W.=V.×Sp. Gr. In order to preserve this simple relationship, it has been found convenient to use in chemistry, for estimating the weight of aëriform substances, another unit called the *crith.* The crith is the weight, *in vacuo*, of one litre of hydrogen gas at 0° centigrade, and with a tension of 76 centimetres. It is equal to 0.09 of a gramme nearly. We may now define the density of a gas as the weight of one litre of the substance in criths, and its specific gravity as a number which shows how many times heavier the aëriform substance is than an equal volume of hydrogen under the same conditions of temperature and pressure. We always estimate the absolute weight of a gas under what we call the standard condition, namely, when the centigrade thermometer marks 0°, and the barometer stands at 76 centimetres. But, in determining the specific gravity of a gas, the comparison with the standard gas may be made at any temperature or pressure, since, as all gases are affected alike by equal changes in these conditions, the relative weights of equal volumes will not be altered by such changes. The subject may be made more clear by the following table:

The specific gravity of a gas shows how many times heavier the aëriform substance is than an equal volume of hydrogen gas under the same conditions of tempera-

ture and pressure. The same number also expresses the weight in criths of one litre of the gas under the standard conditions.

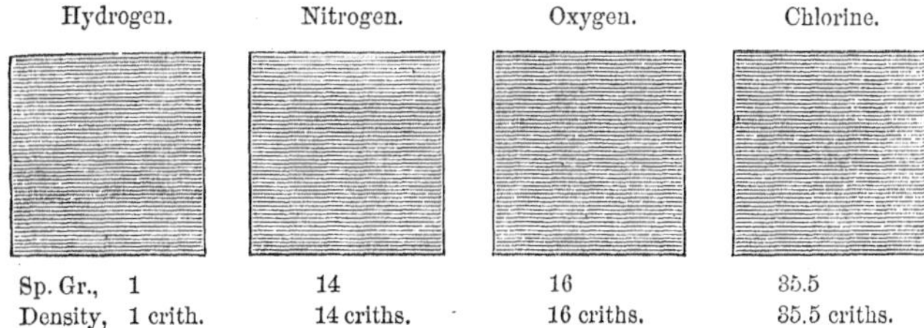

Now we have again W. $=$ V. $\times$ Sp. Gr., only we must remember that W. here stands for a certain number of criths, V. for a certain number of litres, and Sp. Gr. for the specific gravity of the gas referred to hydrogen, a number which also expresses the weight of one litre of the gas in criths.

To return now to the subject of molecular weights. If one litre of hydrogen weighs one crith, and one litre of oxygen sixteen criths, and if both contain the same number of molecules, then each molecule of oxygen must weigh sixteen times as much as each molecule of hydrogen. Or, to put it in another way, represent by n the constant number of molecules, some billion billion, which a litre of each and every gas contains, when under the standard conditions of temperature and pressure. Then the weight of each molecule of hydrogen will be $\frac{1}{n}$ of a crith, and that of each molecule of oxygen $\frac{16}{n}$ of a crith, and evidently

$$\frac{1}{n} : \frac{16}{n} = 1 : 16;$$

that is, again, the weights of the molecules have the same relation to each other as the weights of the equal gas-volumes. Excuse such an obvious demonstration, but it is so important that we should fully grasp this conception that I could not safely pass it by with a few words. It is so constantly the case that the simplest processes of arithmetical reasoning appear obscure when the objects with which they deal are not familiar.

Since, then, a molecule of any gas weighs as much more than a molecule of hydrogen, as a litre of the same gas weighs more than a litre of hydrogen, it is obvious that, if we should select the hydrogen-molecule as the unit of molecular weights, then the number representing the specific gravity of a gas would also express the weight of its molecules in these units. For example, the specific gravity of oxygen gas is 16, that is, a litre of oxygen is sixteen times as heavy as a litre of hydrogen. This being the case, the molecule of oxygen must weigh sixteen times as much as the molecule of hydrogen, and, were the last our unit of molecular weights, the molecule of oxygen gas would weigh 16. So for other aëriform substances. In every case the molecular weight would be represented by the same number as the specific gravity of the gas referred to hydrogen.

Unfortunately, however, for the simplicity of our system, but for reasons which will soon appear, it has been decided to adopt as our unit of molecular weight not the whole hydrogen-molecule, but the half-molecule. Hence, in the system which has been adopted, the molecule of hydrogen weighs 2; the molecule of oxygen, which is sixteen times heavier, 16 times 2, or 32; the molecule of nitrogen, which is fourteen times heavier, 14 times 2, or 28; and, in general, the weight of the molecule of any gas is expressed by a number equal to twice its specific gravity referred to hydrogen. Nothing, then, can be simpler than the finding of the molecular weight of a gas or vapor on this system. We have only to determine the specific gravity of the aëriform substance with reference to hydrogen gas, and double the number thus obtained. The resulting product is the molecular weight required in terms of the unit adopted, namely, the half-molecule of hydrogen. Perhaps there may be some one who, having lost one or more of the steps in the reasoning, wishes to ask the question, Why do you *double* the specific gravity in this method? Let me answer by recapitulating. It all depends on the unit of molecular weights we have adopted. Had we selected the whole of a hydrogen-molecule as our unit, then the number expressing the specific gravity of a gas would also express its molecular weight; but, on account of certain relations of our subject, not yet explained, which make the half-molecule a more convenient unit, we use for the molecular weights a set of numbers twice as large as they would be on what might seem, at first sight, the simpler assumption.

In order to give a still greater definiteness to our conceptions, I propose to call the unit of molecular weight we have adopted a *microcrith*, even at the risk of coining a new word. We already have become familiar with the *crith*, the weight of one litre of hydrogen, and I have now to ask you to accept another unit of weight, the half hydrogen-molecule, which we will call for the future a microcrith. Although a unit of a very different order of magnitude, as its name implies, the microcrith is just as real a weight as the crith or the gramme. We may say, then, that

A molecule of hydrogen weighs 2 microcriths.
" oxygen " 32 "
" nitrogen " 28 "
" chlorine " 71 "

Now, what I am most anxious to impress upon your minds is the truth that, if the molecules, as we believe, are actual pieces of matter, these weights are real magnitudes, and that we have the same knowledge in regard to them that we have, for example, in regard to the weights of the planets. The planets are visible objects. We can examine them with the telescope; and, when we are told Jupiter weighs 320 times as much as the earth, the knowledge seems more real to us than the inference that the oxygen-molecule weighs 32 microcriths. But you must remember that your knowledge of the weight of Jupiter depends as wholly on the law of gravitation as does your knowledge of the weight of the molecules of oxygen on the law of Avogadro. You cannot, directly, weigh either the large or the small mass. Your knowledge in regard to the weight is in both cases inferential, and the only question is as to the truth of the general principle on which your inference is based. This truth admitted, your knowledge in the one case is just as real as it is in the other. Indeed, there is a striking analogy between the two. The units to which the weights are respectively referred are equally beyond the range of our experience only on the opposite sides of the common scale of magnitude; for what more definite idea can we acquire of the weight of the earth than of the molecule of hydrogen, or its half, the microcrith? It is perfectly true that, from the experiments of Maskelyne, Cavendish, and the present Astronomer-Royal of England, we are able to estimate the approximate weight of the earth in pounds, our familiar standard of weight; and so, from the experiments of Sir W. Thompson, we are able to estimate approximately the weight of the hydrogen-molecule, and hence find the value of the microcrith in fractions of the crith or gramme.[1] It is true that the limit of error in the last case is very much larger than in the first, but this difference is one which future investigation will in all probability remove.

I have dwelt thus at length on the definition of molecular weight, because, without a clear conception of this order of magnitudes, we cannot hope to study the philosophy of chemistry with success. Our theory, I grant, may all be wrong, and there may be no such things as molecules; but, then, the philosophy of every science assumes similar fundamental principles, of which the only proof it can offer is a certain harmony with observed facts. So it is with our science. The new chemistry assumes as its fundamental postulate that the magnitudes we call molecules are realities; but this is the only postulate. Grant the postulate, and you will find that all the rest follows as a necessary deduction. Deny it, and the "New Chemistry" can have no meaning for you, and it is not worth your while to pursue the subject further. If, therefore, we would become imbued with the spirit of the new philosophy of chemistry, we must begin by believing in molecules; and, if I have succeeded in setting forth in a clear light the fundamental truth that the molecules of chemistry are definite masses of matter, whose weight can be accurately determined, our time has been well spent.

Before concluding this portion of my subject, it only remains for me to illustrate the two most important practical methods by which the molecular weights of substances are actually determined. It is evident from

[1] According to Thompson, one cubic inch of any perfect gas contains, under standard conditions, 10^{23} molecules. Hence, one litre contains 61×10^{23} molecules and 1 crith $= 122 \times 10^{23}$ microcriths.

what has been said that we can easily find the molecular weight of any substance capable of existing in the state of gas or vapor, by simply determining experimentally the specific gravity of such gas or vapor with reference to hydrogen. Twice the number thus obtained is the molecular weight required in microcriths.

Now, the specific gravity of an aëriform substance is found by dividing the weight of a measured volume of the substance by the weight of an equal volume of hydrogen gas under the same conditions. This simple calculation implies, of course, a knowledge of two quantities: first, the weight of a measured volume of the substance, and, secondly, the weight of an equal volume of hydrogen gas under the same conditions. Of these two weights, the last can always be calculated (by the laws of Mariotte and Charles) from the weight which a cubic decimetre of hydrogen, under the standard conditions, is known to have, namely, 0.0896 gramme or 1 crith; so that the method practically resolves itself into weighing a measured volume of the gas or vapor and observing the temperature and pressure of the substance at the time. There are always at least four quantities to be observed: first, the volume of the gas or vapor; secondly, its weight; thirdly, its temperature; fourthly, its tension; and, lastly, the weight of an equal volume of hydrogen, under the same conditions, is to be calculated from the known data of science.

The most common case that presents itself is that of a substance which, though liquid or even solid at the ordinary temperature of the air, can be readily converted into vapor by a moderate elevation of temperature; such a substance, for example, as alcohol. Now, we can find the weight of a measured volume of such a vapor at an observed temperature and tension in one of two ways, both of which are in general use. In the first process we fill a glass globe of *known size* with the vapor, and *weigh* this measured volume. In the second, we weigh out in a liliputian glass bottle a small quantity of the substance, and, having converted the whole of it into vapor, we measure the volume which it yields.

The first process, devised by Dumas, of Paris, and known by his name, is conducted as follows: We take a glass matrass (a thin glass globe, with a long neck), and, heating the neck in a glass-blower's lamp (as near to the body of the matrass as possible) we draw it out into a capillary tube, three or four inches long. Having first weighed the glass, we introduce into the globe a few table-spoonfuls, we will say, of pure alcohol; and this we can readily do by alternately heating and cooling the vessel. We then mount the globe in a brass frame, and sink it under melted paraffine, but so that the capillary opening shall rise above the surface of the hot liquid. A common iron pot serves to hold the paraffine (Fig. 18), which is heated over a gas-lamp, and a thermometer dipping in the bath enables us to watch the temperature.

Of course, the alcohol is soon volatilized, and the balloon filled with its vapor. The excess escapes through the capillary tube, and, by lighting the jet, we can tell when the vapor in the globe is in equilibrium with the external air, for at that moment the flame will go out. We now, with a blow-pipe, melt the glass around the opening of the capillary tube, and thus hermetically seal up the vapor in the globe. At the same time we note the height of the barometer and the temperature of the bath. The height of the barometer gives us the tension of the vapor in the balloon, because, at the moment of sealing, the tension was equal to the pressure of the air which the barometer directly measures, and the temperature of the vapor must be the same as the temperature of the bath.

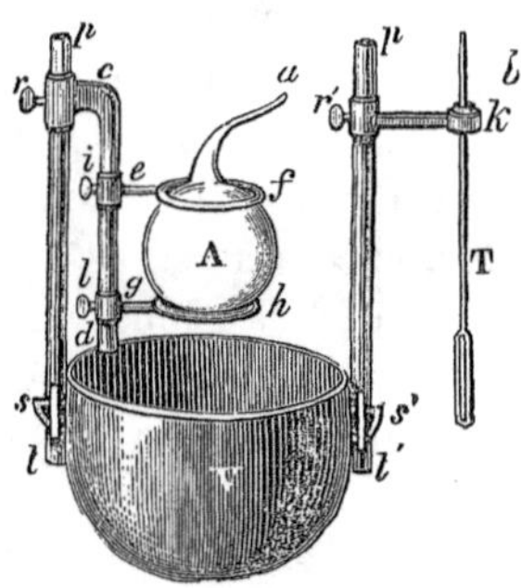

FIG. 18.—Dumas' Method of finding the Specific Gravity of Vapors.

We can now remove the globe, and, after it is cooled and carefully cleaned, weigh it at our leisure. We must remember, however, that the apparent weight of the globe in the balance is not its true weight, because, like a balloon, the globe is buoyed up by the air it displaces, and we must therefore correct the observed weight by adding to it the weight of the air displaced. This correction our knowledge of the weight of air under varying conditions enables us to calculate with the greatest accuracy, assuming, of course, that the volume of the globe is known; and, when, from the weight of the globe thus corrected, we subtract the weight of the glass previously found, the remainder is the weight of alcohol-vapor which just filled the globe at the moment of sealing, and when it had the temperature and pressure we have noted.

Of the four quantities required, we have now observed three, namely, the weight of the vapor, its temperature, and its tension. We also know that its volume was that of the globe when we sealed up its mouth. Since, however, we use a new globe for each determination, we have always to measure its volume, and this, practically, is the last step of the process. The volume is most readily found by filling the globe with water, and weighing. The weight of the water in grammes gives the volume of the globe in cubic centimetres very closely. The globe, moreover, is easily filled, because the condensation of the vapor, on cooling, leaves a partial vacuum in the interior, into which the water rushes with great violence as soon as the tip is broken off under the surface of the liquid. Omitting certain small corrections which it is not best to discuss in this general exposition of the subject, we may, lastly, arrange our calculation in the following form:

Determination of the Molecular weight of Alcohol, by Dumas' Method.

Volume of glass globe..................	500 cubic centimetres.
Temperature at time of closing.........	273° centigrade.
Height of barometer measuring the tension of vapor at time of closing.....	76 centimetres.
Weight of globe and vapor.............	234.29 criths.
Correction for buoyancy, equal to weight of 500 cubic centimetres of air at 0° cent. and 76 centimetres, the temperature and pressure in the balance-case when the globe was weighed...	7.21 "
	241.50 "
Weight of glass........................	230. "
Weight of alcohol-vapor	11.50 "
Weight of 500 cubic centimetres of hydrogen gas at 273°, and 76 c. m. found by calculation, as explained above...........................	0.5 "

$11.50 \div 0.5 = 23$ sp. gr. of alcohol-vapor.

$23 \times 2 = 46$ molecular weight of alcohol.

The second process to which I referred was originally invented by Gay-Lussac, but recently has been very greatly improved by Professor Hofmann, of Berlin. Hofmann's apparatus (Fig. 19) consists of a wide barom-

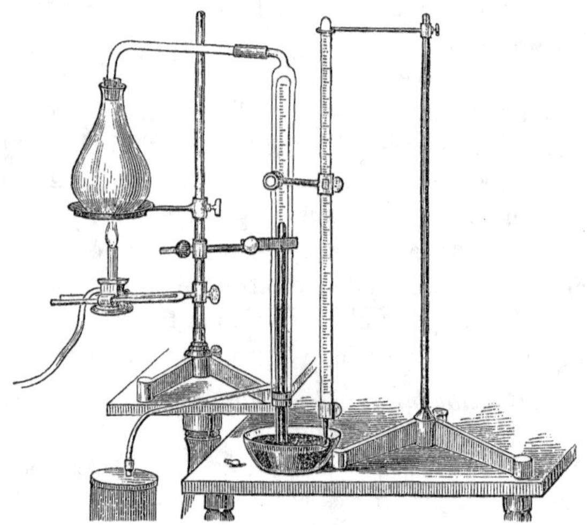

FIG. 19.—Hofmann's Method of finding the Specific Gravity of Vapors.

eter-tube, about a metre long, and graduated into cubic centimetres. This tube is filled with mercury, and inverted over a mercury-cistern, as in the experiment of Torricelli (Fig. 20). The mercury sinks, of course, to the height of about 76 centimetres, leaving a vacuous space at the top of the tube, and into this space is passed up a very small glass-stoppered bottle, containing a few criths of the substance to be experimented on. Around the upper part of the tube is adjusted a somewhat larger tube, also of glass, which serves as a jacket, and through this is passed steam (or the vapor of a liquid boiling at a higher temperature than water), in order to heat the apparatus to a constant and known temperature.

Let us suppose that the substance, whose molecular weight we now wish to find, is common ether. We

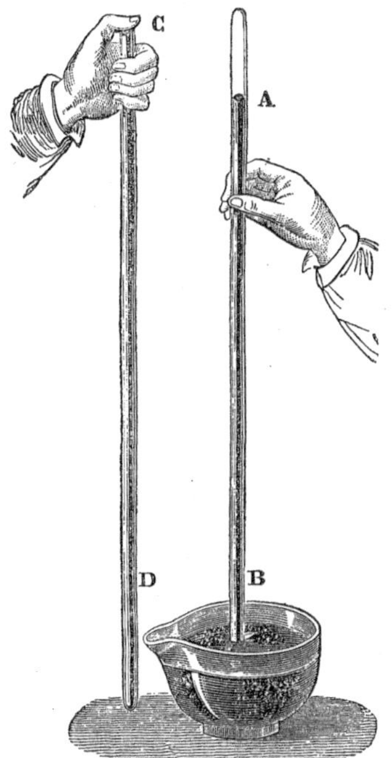

Fig. 20.—Torricelli's Experiment.

begin by weighing our little bottle, first when empty, and then when filled with ether, thus determining, with great accuracy, the weight of the quantity of ether used. With a little dexterity we next pass the bottle under the mercury into the barometer-tube, when it at once rises into the vacuous space. We now pass free steam through the jacket, until we are sure that the temperature of the apparatus is constant at, say,

100° centigrade. The ether, expanding with the heat, soon forces out the glass stopper by which it was confined, and evaporates into the space above the mercury, depressing the column. At first the column oscillates violently, but it soon comes to rest, and we can then read on the graduated scale the volume of the vapor which the weight of ether taken has yielded. This vapor is evidently at the temperature of boiling water, or 100° centigrade; but what is its tension?

The method of measuring the tension will be obvious if you reflect that, in this apparatus, the pressure of the air on the surface of the mercury in the cistern is balanced by the mercury column in the tube and the tension of the vapor pressing on the upper surface of this column. Hence, the height of the column in the tube will be less than that of a true barometer in the neighborhood by just the amount of this tension. In order to find the tension, we have, therefore, only to observe the height of the barometer, and subtract from this the height of the column in our tube, which we must now measure with as much accuracy as possible. Omitting, as in the previous example, a few small corrections, our calculation will now appear thus:

Determination of the Molecular weight of Ether by Gay-Lussac's method, improved by Hofmann.

Weight of ether taken..................	2.539 criths.
Volume of vapor formed................	125 cubic centimetres.
Temperature of vapor..................	100° centigrade.
Height of barometer...................	76 c. m.
Height of column in tube..............	19 c. m.
Tension of vapor......................	57 centimetres.
Weight of 125 cubic centimetres of hydrogen gas at 100° and 57 centimetres, by calculation...............	0.0686 of a crith.

$$2.539 \div 0.0686 = 37 \text{ sp. gr. of ether.}$$

$$37 \times 2 = 74 \text{ molecular weight of ether.}$$

As has been stated, the two methods of determining molecular weight, just described, apply only to those substances which can be readily volatilized by a moderate elevation of temperature. With some slight modifications, the first method may likewise be used for the permanent gases; and, by employing a globe of porcelain, St.-Claire Deville has succeeded in determining, in the same way, the molecular weight of several substances which do not volatilize under a red heat. But a great number of substances cannot be volatilized at all within any manageable limits of temperature, and a still larger number are so readily decomposed by heat as to be incapable of existing in the aëriform condition. The molecular weight of such bodies cannot, of course, be determined by direct weighing. In most cases, however, we are able to infer with considerable certainty the molecular weight of these non-volatile bodies from a knowledge of their composition and other chemical relations; but, nevertheless, there are numerous instances in which the conclusions thus drawn are very questionable, and a great deal of the uncertainty, which still obscures the philosophy of our science, arises from this circumstance.

LECTURE IV.

CHEMICAL COMPOSITION—ANALYSIS AND SYNTHESIS—THE ATOMIC THEORY.

In my previous lectures I have endeavored to give you a clear idea of the meaning which our modern science attaches to the word *molecule.* I must next attempt to convey, as far as I am able, the corresponding conception which the chemist expresses by the word *atom.* The terms molecule and atom are constantly confounded; indeed, have been frequently used as synonymous; but the new chemistry gives to these words wholly different meanings. We have already defined a molecule as the smallest mass into which a substance is capable of being subdivided without changing its chemical nature; but this definition, though precise, does not suggest the whole conception; for the molecule may be regarded from two very different points of view, according as we consider its physical or its chemical relations. To the physicist, the molecules are the points of application of those forces which determine or modify the physical condition of bodies, and he defines molecules as the small particles of matter which, under the influence of these forces, act as units. Or, limiting his regards to those phenomena from which our knowledge of molecular masses is chiefly derived, he may prefer to define molecules as those small particles of bodies which are not subdivided when the state of aggregation is changed by heat, and which move as units under the influence of this agent.

To the chemist, on the other hand, the molecules determine those differences which distinguish substances. Sugar, for example, has the qualities which we associate with that name, because it is an aggregate of molecules which have those qualities. Divide up a lump of sugar as much as you please. The smallest mass that you can recognize still has the qualities of sugar; and so it must be, if you continue the division down to the molecule. The molecule of sugar is simply a very small piece of sugar. Dissolve the sugar in water, and we obtain a far greater degree of subdivision than is possible by mechanical means; a subdivision which, we suppose, extends as far as the molecules. The particles are distributed through a great mass of liquid, and become invisible; still, the qualities of the sugar are preserved; and, on evaporating the water, we recover the sugar in its solid condition; and, according to the chemist, the qualities are preserved, because the molecules of sugar have remained all the while unchanged.

Consider, in the second place, a lump of salt. You do not alter its familiar qualities, however greatly you may subdivide it, and the molecules of salt must have all the saline properties which we associate with this substance. Dissolve the salt in water, and you simply divide the mass into molecules. Convert the salt into vapor, as you readily can, and again you isolate the molecules as before. But, through all these changes, the salt remains salt; it does not lose its savor, because the individuality of the molecules is preserved. So is it with every substance. It is the molecules in which the qualities inhere. Hence the chemist's definition of a molecule: *The smallest particles of a substance in which its qualities inhere, or the smallest particles of a substance which can exist by themselves;* for both definitions are essentially the same.

Hitherto we have only considered molecules as differing from each other in weight, and have learned how to determine their weight; but now we have to regard them as differing in all those qualities which distinguish substances. Considering only the ordinary chemical relations of the two substances, a molecule of sugar differs from a molecule of salt in precisely the same way that a lump of sugar differs from a lump of salt. In a word, what is true of the substance in mass is true of its molecules. Hence it is that, in studying the chemical relations of substances, we may, as a rule, confine our attention to the relations between their molecules, and this very greatly simplifies the problems with which we have to deal; and, in the admirable system of chemical notation, to which I shall hereafter call your attention, the symbol of a substance stands for one molecule, and in using these symbols to represent chemical changes—reactions, as we call them—we always express the reaction as taking place between the individual molecules of the substances concerned.

But, although the molecules are the limit of the physical subdivision of a substance, the chemist carries the subdivision still further; but, then, the parts obtained have no longer the qualities of the original substance, and one or more new substances result. Of course, the chemist cannot, any more than the physicist, experiment on individual molecules. He must experiment on a mass of the substance, and the division of the molecule must be an inference from the phenomena which ensue. Let me call your attention to a few experiments which will illustrate this point:

I crush this lump of sugar in a mortar, and reduce it to what appears to be an impalpable powder, but a microscope will show that the powder consists of grains which are simply smaller lumps, and, in fact, masses of great size, compared with many organisms which are the objects of microscopic investigation. Each one of these grains is sugar, and has all the essential qualities of sugar just as much as the lump. We next pour the powdered sugar into water, in which, as we say, it dissolves; but the solution simply consists in dividing the grains still more, reducing them to molecules, which become spread throughout the mass of the liquid. How are we to go any further than this? Very easily. I take a few more lumps of sugar, and throw them into this heated platinum crucible, when, in an instant, a remarkable change takes place. We have the appearance of flame, and out of the sugar is evolved a mass of loose charcoal. Evidently, this charcoal must have come from the sugar. The crucible is unchanged, and, besides the air, the sugar and platinum were the only substances present. Let me, however, enforce this conclusion by still another experiment, which is even more striking:

Instead of acting on the sugar simply with heat, we will now act upon it with a strong chemical agent called sulphuric acid. For this purpose I have previously prepared about half a pint of very thick syrup, and with this I will now mix three or four times its volume of common oil of vitriol, constantly stirring the mass as my assistant pours in the acid. The syrup at once blackens; soon it begins to swell, and now notice this enormous body of loosely-coherent charcoal which rises from the vessel. Here, again, the charcoal must have been evolved out of the sugar, for the sugar was the only substance common to the two experiments; and, admitting this fact, see to what it leads.

The qualities of sugar inhere in its smallest particles, and must belong to the molecules just as truly as to these lumps. In our experiment the charcoal has been evolved out of a considerable mass of sugar; but the result would have been the same could we experiment on the individual molecules. It is evident, therefore, that the charcoal has been formed out of the sugar-molecules, and that each molecule has contributed its portion to this result. *Now, this charcoal, although so bulky, weighs far less than the sugar.* It could, then, have formed only a part of the mass of the sugar, and only a part of the mass of each molecule. But what has become of the rest of the material? For the present, it must be sufficient to state that careful experimenting has shown that, in this process, another substance is evolved from the sugar besides charcoal, and that this substance is water. Moreover, since the weight of the water, added to that of the charcoal, entirely accounts for the material of the sugar, we conclude that in our experiment the sugar has been resolved solely into charcoal and water. Each molecule, therefore, has been resolved into charcoal and water. In a word, the molecule has been divided. We cannot divide it by any physical means; but we can divide it by chemical means, only we do not obtain thereby two smaller particles of sugar, but a particle of charcoal and a particle of water. Such, then, is the evidence we have that a molecule of sugar can be divided; but the reasoning here used is so important to the validity of our modern chemical philosophy that I must not pass it by with a single example:

One of the substances evolved from the sugar was water. Let us next see whether the molecules of this most familiar substance can be divided. We have already seen to what a wonderful degree of tenuity we can carry the mechanical subdivision of this material. The film of a soap-bubble, just before it bursts, is less than $\frac{4}{1,000,000}$ of an inch in thickness. A square inch of this film would weigh only one $\frac{1}{1,000}$ of a grain. Now, the unaided eye can easily distinguish the $\frac{1}{100}$ of an inch in length, or $\frac{1}{10,000}$ of a square inch of area or a quantity of water in that film, weighing only $\frac{1}{10,000,000}$ of a grain. But a still greater subdivision than this is possible, for, as we now know, when water is converted into vapor, the liquid mass breaks up into small particles of wonderful tenuity, which we call molecules, and by expanding the vapor we can separate these molecules to an indefinite extent. We cannot, it is true, follow this subdivision with the eye, but we can discern it with the intellect; and, furthermore, by determining the specific gravity of aqueous vapor with reference to hydrogen gas, we can very easily find the weight of the aqueous molecules, and we thus know that a molecule of water weighs eighteen microcriths. By physical processes we cannot carry the subdivision any further. The smallest mass of water of which we have any knowledge weighs eighteen microcriths; but we can divide the molecule chemically, as the following experiment will prove:

In order to show you the decomposition of water by an electrical current, I have projected on the screen the magnified image of a glass cell containing a small quantity of this familiar liquid, acidulated, however (with sulphuric acid), in order to make it a conductor of electricity. Connected with the cell is what must be known to all of my audience as a voltaic battery. The conducting wires from the end plates of this combination terminate in the two strips of platinum, which you see projected on the screen. As soon as the connections are made, or, to use the technical phrase, as soon as the circuit is closed, an electric current flows through the water in the cell, passing from one of these poles to the other. The effect of this current is visible. Bubbles of gas collect upon the platinum strips, and, as soon as they attain sufficient size, rise to the surface of the water, and this evolution of gas will go on so long as the electric current continues to flow. The gases evolved at the two poles are wholly different substances, and, in order to exhibit to you their characteristic qualities, I have prepared a second experiment:

Standing on the table is a decomposing cell similar to the last, but very much larger, and so constructed that the two gases are collected as they rise from the poles, and conducted apart into these two glass bells. A very powerful electric current has been passing through the water in the cell since the beginning of the lecture, and already the bells are filled with the two aëriform products. Both are invisible, but notice that the gas we have collected in the right-hand bell takes fire and burns with a pale and barely luminous flame. Here we have a very large bell full of the same gas, and on lighting this I think the flame will be visible to all. Every one must have recognized this material.

It is a well-known substance, which we call hydrogen. It is one of the very few substances which we only know in the aëriform condition. It is, moreover, the lightest form of matter known. A cubic yard of air at the temperature of this room (77° Fahr.) weighs, in round numbers, two pounds, while a cubic yard of hydrogen weighs only two and a half ounces. These rubber balloons, which are such familiar toys, illustrate very forcibly the wonderful lightness of this singular form of matter.

Let us turn now to the gas in the left-hand bell, and we shall find that it differs most strikingly from the other, and in no respect is the difference more marked than in the weight. This gas is sixteen times heavier than hydrogen, that is, the difference between the density of the two is almost as great as that between iron and cork, and yet these invisible forms of matter are so intangible that it is difficult even for the chemist to appreciate this difference. Bringing now a lighted candle near the open mouth of the bell, you see that the gas will not burn; but notice that, as I lower the candle into the bell, the wax burns in the gas far more brilliantly than in air. Observe, also, that this smouldering slow-match bursts into flame when immersed in the same medium. Evidently it supports combustion with great vigor, and, in order to illustrate this point still more strikingly, I will introduce into another bell of the same gas a spiral of watch-spring tipped, like a match, with a little sulphur, first setting fire to the sulphur. . . . See! the iron burns as readily as tinder, and far more brilliantly. We are dealing, in fact, with oxygen, the same gas which is found all around us in the earth's atmosphere—only, in our atmosphere the oxygen is mixed with four times its volume of an inert gas called nitrogen, while as evolved from the water in our experiment it is perfectly pure.

It is evident, then, that in this experiment two

new substances are evolved, and the question arises, Whence do they come? If we examine carefully the conditions of the experiment we should find that, of all the substances present, the only one which underwent any permanent change was the water. The weight of the platinum poles, for example, remains unchanged, but the weight of the water is diminished in exact proportion to the amount of gas evolved. These aëriform substances are then educed from the material of the water. Moreover, it has also been proved that the water is *completely* resolved into these gases. The electric current is merely a form of energy, and, of course, can neither add nor remove ponderable material, and the weight of oxygen and hydrogen formed is exactly equal to the weight of water lost. As we say in chemistry, the electric current analyzes the water, and these gases are its sole constituents.

Let me now call your attention to another fact connected with the process we are studying; and, in order that you may observe the fact for yourselves, I will repeat the experiment with still a third apparatus, so constructed that we can measure the volumes of the two gases which are formed. I have placed the apparatus in front of my lantern so that I can project on the screen a magnified image of the graduated tubes in which the gases are collected.

You notice that the volume of one is *twice* as large as that of the other, and this ratio is found to hold exactly when we make the experiment with the very greatest accuracy. The larger volume is hydrogen, the lesser oxygen. But oxygen, as I have said, is *sixteen* times as heavy as hydrogen. Hence, there is *eight* times as much material in the half-volume of oxygen as in the whole volume of hydrogen, or, in other words, when water is decomposed by electrolysis, there is eight times as much oxygen produced as hydrogen.

We regard, then, this experiment as establishing, beyond all controversy, the fact that water is composed of oxygen and hydrogen gases in the proportions of

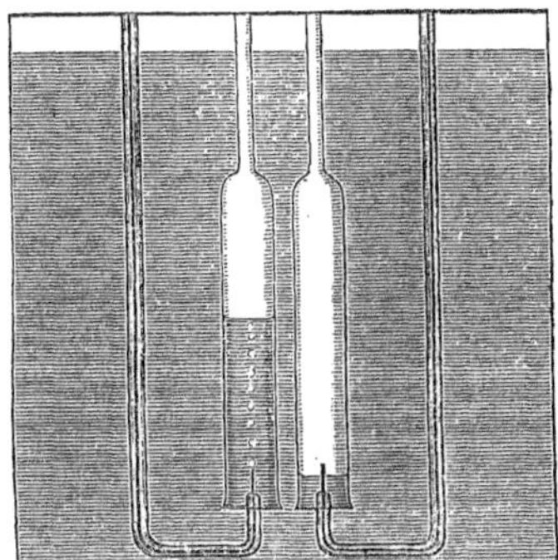

FIG. 21.—Decomposition of Water by Galvanism.

eight to one, or, in other words, that in every nine parts of water there are eight parts of oxygen and one part of hydrogen. But, if this is true, it must be true of the smallest mass of water as well as of the largest. It must be true, then, of the molecule of water. Now, one molecule of water weighs 18 microcriths. Hence, of those 18 microcriths, one-ninth, or two microcriths, must consist of hydrogen, and eight-ninths, or 16 microcriths, must consist of oxygen. Please notice that this is a result to which our experiment directly leads, and is as much a fixed truth as any results of observation. Unless our whole science is in error, and Avogadro's law a delusion, then it is an established fact that the molecule of water weighs 18 microcriths, and equally certain that this molecule consists of 16 microcriths of oxygen, and of 2 microcriths of hydrogen. Moreover, it is also evident that, when we analyze water, as in this experiment, the molecules are divided, and that, from the material thus obtained are formed the molecules of the two aëriform substances which are the products of the process. As yet I advance no theory as regards the nature of this process, or of the condition in which the two substances exist in the molecule of water. I am only dealing with the bare fact that they are evolved out of the molecule, and that the molecule is thus divided. There are a great many other chemical processes by which water may be analyzed, and the result is in all cases precisely the same, namely, that from every nine parts of water there are obtained eight parts of oxygen and one of hydrogen. Of course this concurrence of testimony is very valuable, but we need not go beyond this simple experiment to establish the truth we have enunciated, and our experiment has this great advantage for the present purpose: There is nothing to complicate the process, and you can be almost said to see that the oxygen and hydrogen come from the water and from that alone.

Such illustrations might be very greatly multiplied, but the two we have selected are sufficient to show how the chemist is able to divide the molecule, and that this division is always attended with the destruction of the original substance, and the evolution from it of wholly different substances. We are now, then, prepared to classify the various changes which we observe in Nature, and define them according to the terms of the molecular theory.

There are many changes in which the identity of the substance remains unimpaired, although the external form may be greatly altered, and in some cases even new qualities acquired. Thus, a bar of iron may be drawn out into wire finer than the finest hair, may be rolled out into leaves of exceeding tenuity, may become magnetized, and thus acquire a remarkable power of attracting other masses of iron; but all this time the material remains unchanged, and would be recognized by every one as iron. Such changes as these are called physical changes. They are necessarily attended with very great changes in the relative position of the molecules, or even in their condition; but the molecules remain undivided, and retain throughout their integrity. But there is another class of changes, whose very essence consists in the conversion of the substances involved into new substances. Coal and wood burn, and are thereby converted into those aëriform substances which we designate collectively under the name of smoke. Iron rusts, and changes into a yellowish-red powder. Out of white, sweet, soluble sugar comes this porous mass of black charcoal, and out of water come oxygen and hydrogen gases. Such changes as these are called chemical changes. They are caused by a change in the old molecules. New ones are formed, and hence new substances are formed.

In some cases the old molecules are divided into parts of a different nature. Thus, the molecules of sugar are divided into masses of charcoal and water, and the molecules of water again are divided into particles of oxygen and hydrogen. In such cases, we say that the substance is decomposed into its constituent parts. In other cases, the old molecules attach to themselves more material, and new molecules, of greater weight, result, and we then say that the substance has combined with another, as the coal with oxygen in the process of burning, and the iron with oxygen in the process of rusting. The first class of changes we call analysis, the second, synthesis. The evidence of analysis is that each product of the change weighs less than the substance from which it was evolved. The evidence of synthesis is that the total product weighs more than the original substance.

The oxygen and hydrogen gases, each apart, weigh less than the water from which they were formed, and the fact that the sum of their weights is exactly equal to that of the water, proves that they are the only products of the change, and that water is composed of these substances, and of these alone. The gas we call carbonic dioxide, which is the only product of the burning of pure coal, weighs more than the coal, and, since this excess of weight is exactly equal to that of the oxygen consumed in the burning, we conclude that, in this process, the coal has combined with oxygen, and that the carbonic dioxide is a compound of these two substances.

Thus arise our scientific conceptions of combination and decomposition, of synthesis and analysis. When we say that sugar is composed of charcoal and water, we mean merely that these two substances may be evolved from sugar; and the evidence that they are the only constituents of sugar is that the sum of the weights of the two products equals the weight of the sugar. When we say that water is composed of oxygen and hydrogen, we merely mean that these two substances may be educed from water, and that, as before, the weight of the two products exactly equals the weight of the water. When we say that carbonic dioxide is composed of charcoal and oxygen, our assertion is based on the fact that, in the process of burning, the oxygen gas appears to absorb charcoal, and that the resulting gas weighs more than the oxygen by the exact weight of the charcoal consumed. In the first two cases, the proof of the composition is analytical, in the third synthetical. In many cases we have both modes of proof. Thus, we can decompose water into oxygen and hydrogen gases, and show that the weight of the products is exactly the same as that of the water which has disappeared. We can also combine hydrogen with oxygen, and show that the weight of water formed is exactly equal to that of the two gases consumed.

Notice the important part which the weight of the substances concerned in our processes plays in this reasoning. That water consists of oxygen and hydrogen, and of nothing else, is a conclusion based on the fact that the weight of the substance has been found equal to that of its assumed constituents. Of course the reasoning implies the truth of the principle that increase of weight always indicates increase of material, and diminution of weight diminution of material, or, in other words, that the weight of a body is proportional to the amount of material it contains. But this principle, so obvious now, is by no means, as might at first appear, self-evident, and it is only comparatively recently that it has become an accepted principle of science. It was never fully enunciated before Newton, and, although his master-mind was able to establish the foundations of astronomy on this basis two centuries ago, it is only comparatively recently that the principle has been fully accepted in chemistry. For years after Newton, the chemists believed universally in a kind of matter called phlogiston, which not only could be removed from a substance without diminishing its weight, but whose subtraction actually added to the weight. It is the great merit of Lavoisier that he clearly conceived of this principle, and insisted on its application in chemistry. He was the first to see clearly that, in every chemical process, increase of weight means increase of material, and loss of weight loss of material. Iron, in rusting, gains in weight. Hence, said Lavoisier, it has combined with some material. No, said the defenders of the phlogiston theory, such men as Cavendish, Priestley, and Scheele, it has only lost phlogiston. You are making too much of this matter of weight. Phlogiston differs from your gross forms of matter in that it is specifically light, and, when taken from a body, increases its weight. We smile at this idea, and we find it difficult to believe that these men, the first scientific minds of their age, could believe in such absurdity. But we must remember that the idea did not originate with them. It was a part of the old Greek philosophy, and from the pages of Aristotle was taught in every school of Europe until within two hundred years; and, even in our own time, we still hear of *imponderable* agents. Text-books of science are used in some of our schools which refer the phenomena of heat and electricity to attenuated forms of matter, that can be added to or subtracted from bodies without altering their weight. Such facts should teach us, not that we are so much wiser than our fathers, but that our familiar ideas of the composition of matter are not such simple deductions from the phenomena of Nature as they appear to us; and this discussion of the evidence, on which these conclusions are based, is therefore by no means superfluous.

As the result of this discussion let us bear in mind that, when we say that water is composed of oxygen and hydrogen, we mean no more than this, that, by various chemical processes, these two substances can be produced from water, and that the weight of the two products always equals the weight of the water employed in the process; or, on the other hand, that water may be produced by the combination of oxygen with hydrogen, and that the weight of the water thus formed is equal to the sum of the weights of the two gases. We cannot say that water consists of hydrogen and oxygen, in the same sense that bread consists of flour, or syrup of sugar, and mortar of lime. We must be very careful not to transfer our ideas of composition, drawn chiefly from the mixtures we use in common life, directly to chemistry. In these mixtures the product partakes, to a greater or less degree, of the character of its constituents, which can be recognized essentially unchanged in the new material, but, in all instances of true chemical union and decomposition, the qualities of the substances concerned in the process entirely disappear, and wholly different substances, with new qualities, appear in their place. Prior to experience, no one could suspect that two aëriform substances like oxygen and hydrogen could be obtained from water, and the discovery of the fact, near the beginning of this century, marks an era in the history of science. And, even now, familiar as it is, this truth stands out as one of the most remarkable facts of Nature. Moreover, the wonder becomes still greater when we learn that water yields 1,800 times its volume of the two gases, and that these gases retain their aëriform condition so persistently that no power has been able to reduce them to the liquid condition; and still more the wonder grows, when we learn further that the amount of energy required to decompose a pound of water into its constituent gases would be adequate to raise a weight of 5,314,200 pounds one foot high; and that, when these gases unite and the water is reproduced, this energy again becomes active. Two experiments will enforce the truth of this statement:

For the first, I have mixed together in this rubber bag oxygen and hydrogen in the exact proportions in which they unite to form water, and, with the gas, I will now blow up into froth the soap-suds contained in this iron mortar—thus confining the gas only by the thinnest possible envelope. I will now ask my assistant to inflame the mixture with his lighted taper, when a deafening explosion announces to us that the chemical union has taken place. But what has been the occasion of the development of such tremendous energy? The formation of a single drop of water, so small that you could hold it on the point of a needle.

For the second experiment I will burn the same gas-mixture at a jet, and show you how great is the intensity of the heat which may be thus developed. This apparatus is the well-known compound blow-pipe invented by our countryman Dr. Hare. The oxygen and hydrogen flow through rubber hose from separate gas-holders into a very small chamber, where they mix before issuing from the jet. The same chemical union takes place here as before; the same product (water) is formed; the same amount of energy is developed; but, under these different conditions, the explosive gas burns with a quiet flame as it is gradually supplied from the jet, and the energy, instead of being expended in driving back the air, and thus determining that violent commotion in the atmosphere which caused the noise, is here manifested wholly as heat. And see how intense the heat is! . . . It is a steel file which is burning with such rapidity in this flame. As I have already told you, heat is only a mode of energy, and, like any other manifestation of power, may be measured in foot-pounds. Hence, this brilliant experiment is an apt illustration of the amount of energy developed in the production of water. In witnessing the magnitude of the effects, we are surprised, as before, by the apparent inadequacy of the cause; for the amount of water, whose production was the occasion of all this display of power, is only a few drops.

Who could believe that such power was concealed in the familiar liquid which is so intimately connected with our daily life? Between the qualities of water and the qualities of these gases there is not the most distant resemblance. When the water is decomposed, the qualities of the water are wholly lost in the qualities of the two gases produced from it, and a certain amount of energy is absorbed. When the water is formed, the qualities of oxygen and hydrogen are wholly merged in those of the resulting liquid, while the same amount of energy is set free. Whether the oxygen and hydrogen exist, as such, in the water, or whether they are produced by some unknown and unconceived transformation of its substance, is a question about which we may speculate, but in regard to which we have no knowledge. All we know is, that the change of water into the two gases or of the two gases into water is attended with no change of weight, and hence we conclude that in the change the material is preserved, or, in other words, that water and the gases are the same material in different forms.

Now, the only theory which has as yet succeeded in giving an intelligible explanation of the facts, assumes that hydrogen and oxygen do exist as such in water, preserving each its individuality; that each molecule of water consists of three particles, two of hydrogen and one of oxygen; that, when the water is decomposed, the molecules are broken up, and that then the oxygen particles associate themselves together to form molecules of oxygen gas, and the hydrogen particles to form molecules of hydrogen gas; that, on the other hand, when the gases recombine, the reverse takes place, each particle of oxygen uniting to itself two particles of hydrogen to form a molecule of water.

These parts of molecules (these particles, into which the molecules break up under various chemical processes) are what we call atoms, and this theory is the famous atomic theory, which has played such a prominent part in modern chemistry. We shall find, as we proceed, that there is very strong evidence in its support. Indeed, without it a large part of the modern science would be wholly unintelligible; and, were I to confine my regards to purely chemical facts, I should regard the evidence in its favor as overwhelming. Still, I must confess that I am rather drawn to that view of Nature which has favor with many of the most eminent physicists of the present time, and which sees in the cosmos, besides mind, only two essentially distinct beings, namely, matter and energy, which regards all matter as one and all energy as one, and which refers the qualities of substances to the affections of the one substratum, modified by the varying play of forces. According to this view, the molecules of water are perfectly homogeneous, and the change, which takes place when water is decomposed, does not consist in the separation from its molecules of preexisting particles, but in imparting to the same material other affections.

I know that this language is very vague, but it is

no more vague than the idea it attempts to embody. Still, vague as it is, no one who has followed modern physical discussions can doubt that the tendency of physical thought is to refer the differences of substances to a dynamical cause. Nevertheless, as I said before, the atomic theory is the only one which, as yet, has given an intelligible explanation of the facts of modern chemistry, and I shall next proceed to develop its fundamental principles. I wish, however, before I begin, to declare my belief that the atomic theory, beautiful and consistent as it appears, is only a temporary expedient for representing the facts of chemistry to the mind. Although in the present state of the science it gives absolutely essential aid both to investigation and study, I have the conviction that it is a temporary scaffolding around the imperfect building, which will be removed as soon as its usefulness is passed. I have been called a blind partisan of the atomic theory, but, after this disclaimer, you will understand me when, during the remainder of this course of lectures, I shall endeavor to present its principles as forcibly as I can.

LECTURE V.

ELEMENTARY SUBSTANCES AND COMBINING PROPORTIONS.

In my last lecture I stated that in a chemical compound the qualities of the constituents are wholly merged in those of the product, and that this circumstance distinguishes a true compound from a mechanical mixture in which the qualities of each ingredient are to a greater or less extent preserved. This distinction is one of very great importance in chemistry, and I will begin my lecture this evening by asking your attention to a simple experiment, which will recall the principal points of our discussion at the last lecture and at the same time illustrate still other aspects of this important subject.

I have prepared a mixture of finely-divided iron (iron reduced by hydrogen) and flowers of sulphur. The two powders have been rubbed together in a mortar until the mass appears perfectly homogeneous and it is impossible with the unaided eye to distinguish the grains of either substance, and yet nothing is easier than to show that both are here wholly unchanged.

For this purpose I will, in the first place, pour upon a portion of the powder some of this colorless liquid called sulphide of carbon, which dissolves sulphur with great eagerness. After shaking the two together we find left on the bottom of our glass beaker a quantity of a black powder, which, as the magnet shows at once, is iron. In the second place I will stir up another portion of the mixture with alcohol, using this liquid to hold the powder in suspension so that I can pick out the grains of iron with a magnet. Using this bar-magnet as a stirring-rod, I can thus readily wash out the sulphur from the iron which adheres to the magnet, and we recognize at once the yellow color as the particles of sulphur settle to the bottom of the jar.

Having shown you now that both iron and sulphur are here present, with their qualities wholly unaltered, I will next take a third portion of the powder, and, having made with it a small conical heap, apply a lighted match to the apex of the cone. A glow at once spreads through the whole mass, which is an evidence to me that a chemical change has taken place, and in that change the sulphur and iron have disappeared. The mass has somewhat caked together, but we can easily pulverize it again, and our product is then a black powder not differing very greatly in external appearance from the original material. But from this black powder the sulphide of carbon can dissolve no sulphur, and the magnet can remove no iron.

The qualities both of the iron and the sulphur have disappeared, and those of a new substance we call sulphide of iron have taken their place, and the only evidence we have that the material of the sulphur and the material of the iron are still here is the weight of the sulphide of iron, which is exactly equal to that of the sulphur and iron combined. So long as the sulphide of iron remains sulphide of iron, no scrutiny can detect in it either sulphur or iron, and we must have recourse to other chemical processes in order to reproduce these substances. In old times, before men had clearly conceived that weight is the measure of material, and that, as thus measured, no material is ever lost, it was supposed that in such experiments as this the substances involved underwent a mysterious transformation; the essence of matter, whatever it might be, changing its dress, and appearing in a new garb; and men reasoned, "If such transformations as these are possible, why not any others?" and hence centuries were wasted in vain attempts to transform the baser metals into gold. Our present convictions that such transmutation is impossible are based on the knowledge we have obtained by following to its legitimate consequences the great principle established by Newton: when the weight remains, we are persuaded that the material remains. The weight of the sulphide of iron is exactly equal to that of the sulphur and iron combined. Hence we conclude that every atom of the iron and every atom of the sulphur still remain in our product, the only difference being that, whereas, previously, the atoms of the sulphur were associated together to form molecules of sulphur, and those of the iron to form molecules of iron, they are now associated with each other to form molecules of sulphide of iron.

According to our atomic theory, then, in one sense at least, chemical combination is only a mixture of a finer degree. If we place on the stage of a powerful microscope a portion of the powder with which we have just been experimenting, we can distinguish the grains of sulphur and those of iron, side by side; and so, according to our theory, if we could make microscopes powerful enough, we should see in the sulphide of iron the atoms of its two constituents. But, although, in this one respect, our modern chemistry regards combination as merely a more intimate mixture, yet it recognizes a very great difference between these two classes of products indicated by a most remarkable characteristic, to which I have next to direct your attention.

Chemical combination always takes place in certain definite proportions, either by weight or measure. Thus we may mix together sulphur and iron in any proportion we choose, but when, on heating, combination takes place, 56 grains of iron combine with just 32 grains of sulphur; and, if there is an excess of one or the other substance, that excess remains uncombined. If there is an excess of sulphur, there remains so much free sulphur, which we can dissolve out with sulphide of carbon; and, if there is an excess of iron, there remains so much metallic iron, which we can separate with a magnet. So is it, also, in the combination of oxygen with hydrogen to form water. Eight grains of oxygen combine with exactly one grain of hydrogen, and any excess of either gas remains unchanged, and in all cases of chemical combination and decomposition similar definite proportions are preserved between the weight of the several constituents, which unite to form the compound, or result from its decomposition.

It is an obvious explanation of this law of definite proportions that the small particles or atoms between which the union is assumed to take place, have a definite weight; in other words, are definite masses of matter. Now, the atomic theory supposes, in the combination of sulphur and iron, for example, that the two materials break up into atoms; that an atom of iron unites with an atom of sulphur to form a molecule of sulphide of iron, and that the union takes place in the proportion by weight of 56 to 32, simply because these numbers represent the relative weight of the two sorts of atoms (the atoms of the same substance being all alike, and all having the same size and weight). In the case of water, for reasons which will hereafter appear, it supposes that two atoms of hydrogen combine with one atom of oxygen to form a molecule of water, and, since each atom of oxygen weighs sixteen times as much as an atom of hydrogen, the two substances must combine in the proportion of 2 : 16, or 1 : 8, as stated above.

The principle we have been discussing is known in chemistry as the law of definite proportion. It was first clearly stated by Wenzel and Richter, in 1777, and the atomic theory, although itself as old as philosophy, was first applied to the explanation of the law by the English chemist Dalton, in 1807. Subsequent discoveries have greatly tended to confirm this theory, but, before we can appreciate their bearing on our subject, we must endeavor to grasp another of the elementary conceptions of our science. As in previous cases, I shall not content myself with stating the truth, but endeavor to show how it is deduced from observation.

The study of chemistry has revealed a remarkable class of substances, from no one of which a second substance has ever been produced, by any chemical process, which weighs less than the original substance. Let me illustrate what I mean by a few experiments:

The white powder which is counterpoised on the pan of this balance is called sulphocyanide of mercury, and has been used in the preparation of a toy called Pharaoh's serpent. You have all probably seen the experiment, but perhaps have not observed the feature to which I wish to call your attention. As in the previous experiment, I have made with the powder a small conical heap, and I will now apply the flame of a match to the apex of the cone. The mass takes fire and burns, but, so far from its being consumed, there rolls up from it a great body of stuff whose singular shape suggested the name of the experiment.

It is certainly a most remarkable chemical change; for, from a small amount of white powder, we have produced this great volume of brown material. Moreover, the conditions of the experiment are such that it is evident that the material must have been formed from the white powder. The only other substance present is the atmospheric air, which, although it plays an important part in the change, could not have yielded this singular product. Notice, now, that the product, voluminous as it is, weighs less than the original substance. This is the feature of the experiment to which I wish especially to direct your attention, and the inference to be drawn from it is obvious. The sulphocyanide of mercury has been decomposed, and the material of this brown mass was formerly a part of the material of this substance.

Allow me next to recall to your minds the experiments we made in a previous lecture with sugar. In these experiments the sugar was converted into charcoal, and the conditions were such that the charcoal must have come from the sugar, and from nothing else. Now, since the charcoal weighed less than the sugar, it was evident that the material of the charcoal was a part of the material of sugar, or, in other words, that one of the constituents of sugar is charcoal. As I then stated, charcoal was not the only product of those chemical changes. Water was also produced, and under such conditions that the material of the water must have come from the material of sugar, and from that alone. Hence, we feel justified in concluding that a part of the material of sugar is water; and finding, further, that the weight of the charcoal and water together is equal to that of the sugar, we also conclude that the material of sugar consists of charcoal and water, and of these substances only.

So, also, in the experiment of decomposing water by an electrical current, it is evident that the hydrogen gas produced comes from the water, and, as the hydrogen obtained weighs far less than the water consumed, we conclude that a part of the material of water is hydrogen. For the same reasons we conclude that a part of the material of water is oxygen; and, lastly, since the weight of the oxygen and hydrogen together just equals the weight of the water, we conclude that the material of water consists *wholly* of hydrogen and oxygen. Let me ask your attention now to still another experiment:

I have counterpoised on the pan of a second balance a few grammes of that same finely-pulverized iron which we have already used in this lecture. In this condition metallic iron burns in the air with the greatest readiness. We need only touch the powder with a lighted match when a glow spreads through the mass as through tinder. Notice that the conditions of the experiment are such that no substances can concur in the change except iron and air. As the result of the change a new substance is produced, just as in the other cases, and this substance we call oxide of iron. Is, then, this new substance a part of the material of iron, in the same sense that oxygen is a part of the material of water? The only circumstance which points to a different conclusion is what the balance indicates. The iron has increased in weight, proving that material has been added to it, and not taken from it; and, as you all know, the iron, in burning, has combined with the oxygen of the air. Oxygen, then, is the material which has been added.

This experiment illustrates a most remarkable truth in regard to the substance we call iron. By various chemical processes we can produce from the metal hundreds of different substances, but, in all cases, the conditions of the experiment, and the relative weight of the products, prove that material has been added to the iron, and not taken from it. By no chemical process whatever can we obtain from iron a substance weighing less than the metal used in its production. In a word, we can *extract* from iron nothing but iron.

Now, there are sixty-three (possibly sixty-five) different substances of which this same thing can be said. From no one of these substances have we been able to extract any material save only the substance itself. We are able to convert them into thousands on thousands of other substances; but, in all cases, the relative weight of the products proves that material has been added to, not taken from, the original mass. To use the ordinary language of science, we have not been able to decompose these substances, and they are distinguished in chemistry as elementary substances.

These substances are frequently called chemical elements, but our modern chemistry does not attach to this term the idea that these substances are primordial principles, or self-existing essences, out of which the universe has been fashioned. Such ideas were associated with the word *element* in the old Greek philosophy, and have been frequently defended in modern times; and, so far as the words element and element-

List of Elementary Substances.

Aluminum,	Al,	27.5	Mercury,	Hg,	200
Antimony,	Sb,	120.3	Molybdenum,	Mo,	92
Arsenic,	As,	75	Nickel,	Ni,	58
Barium,	Ba,	137	Nitrogen,	N,	14
Bismuth,	Bi,	203	Osmium,	Os,	199.2
Boron,	B,	11	Oxygen,	O,	16
Bromine,	Br,	80	Palladium,	Pd,	106.6
Cadmium,	Cd,	112	Phosphorus,	P,	31
Cæsium,	Cs,	133	Platinum,	Pt,	197.4
Calcium,	Ca,	40	Potassium,	K,	39.1
Carbon,	C,	12	Rhodium,	Rh,	104.4
Cerium,	Ce,	92	Rubidium,	Rb,	85.4
Chlorine,	Cl,	35.5	Ruthenium,	Ru,	104.4
Chromium,	Cr,	52.2	Selenium,	Se,	79.4
Cobalt,	Co,	58	Silicon,	Si,	28
Columbium,	Cb,	94	Silver,	Ag,	108
Copper,	Cu,	63.4	Sodium,	Na,	23
Didymium,	D,	95	Strontium,	Sr,	87.5
Erbium,	E,	112.6	Sulphur,	S,	32
Fluorine,	F,	19	Tantalum,	Ta,	182
Glucinum,	Gl,	4.7	Tellurium,	Te,	128
Gold,	Au,	197	Thallium,	Tl,	204
Hydrogen,	H,	1	Thorium,	Th,	231.4
Indium,	In,	75.6	Tin,	Sn,	116
Iodine,	I,	127	Titanium,	Ti,	50
Iridium,	Ir,	198	Tungsten,	W,	184
Iron,	Fe,	56	Uranium,	Ur,	118.8
Lanthanum,	La,	93.6	Vanadium,	Va,	51.37
Lead,	Pb,	207	Yttrium,	Yt,	61.7
Lithium,	Li,	7	Zinc,	Zn,	65
Magnesium,	Mg,	24	Zirconium,	Zr,	89.6
Manganese,	Mn,	55			

ary suggest such ideas, they are unfortunate terms. Experimental science, which deals only with legitimate deductions from the facts of observation, has nothing to do with any kind of essences except those which it can see, smell, or taste. It leaves all others to the metaphysicians. It knows no difference between elementary substances and any other class of substances, except the one already pointed out. No one can distinguish an elementary substance by any external signs. Sulphur and charcoal are elementary substances, chalk and flint are compound substances; but who would know the difference? And, seventy-five years ago, men did not know that there was any difference. Modern chemistry has shown, by a process of reasoning precisely similar to that which we have discussed, that out of the material of chalk we can obtain a metal called calcium, and out of flint a combustible substance called silicon; while, from the material of charcoal or sulphur, we can educe no product but the same charcoal or sulphur again. Hence, we say that the first are compound substances, and the last elementary; but, were a process discovered to-morrow by which a new substance was produced from the material of sulphur, we should hail at once the discovery of a new element, and sulphur would be banished forever from the list of elementary substances. Yet the qualities of sulphur would not be changed thereby. It would still be used for making sulphuric acid and bleaching old bonnets, as if nothing had happened. All this may seem very trivial, but there is no idea more common, or of which it is more difficult to disabuse the mind of a beginner in the study of chemistry, than the notion that there is something peculiar or unreal about what is called a chemical element; and the conception that an element is a definite substance, like any other substance, is usually the beginning of clear ideas on the subject. I hope I have been able to make this truth prominent, and also to impress the further truth that all our knowledge of the composition of matter is based on the fundamental principle that weight is the true measure of quantity of material, which is simply the first postulate of the law of gravitation. This great law of Newton is thus the basis of modern chemistry as much as it is of modern astronomy.

We are now prepared to accept intelligently the following general propositions: 1. That all substances may be resolved by chemical processes into one or more of the sixty-three elementary substances; 2. That all substances not themselves elementary may be regarded as formed by the union of two or more elementary substances. Of course, the second is merely the reverse of the first, and is implied by it; but the two represent the two methods of proving the constitution of substances, which we have called analysis and synthesis. Of these the analytical proof alone is universally possible. In by far the larger number of cases, however, we are also able to effect the synthesis of substances by uniting the elements of which they consist, but there is still a considerable number of substances which have never been produced in this way.

Having acquired the conception of an elementary substance, and of its chemical relations, we can now give to the law of definite proportions a more precise statement. As I have already said, the law is universal. It applies to all kinds of chemical changes, and to all classes of substances, elementary as well as compound. But elementary substances are only susceptible of that class of changes we have called synthetical. They can combine with each other, but they cannot be resolved into other substances. Hence all the information in regard to them, which the law, as thus far enunciated, gives us, is that, when they combine, the union takes place in definite proportions by weight or volume. But this is not all the truth. There is a law governing the definite proportions, and the proportions of the different elementary substances which unite to form the various known compounds are so related that it is possible to find for each element a number, such, that, in regard to the several numbers, it may be said that the elements always combine in the proportion by weight of these numbers or of some simple multiples of these numbers. This supplement to the law of definite proportions is known as the *law of multiple proportions*; but, if we accept the atomic theory, both laws are merely necessary consequences of the constitution of matter which this theory assumes to exist. Let us, in the first place, understand fully the facts, and we shall then be prepared to consider their bearing on our theory.

In the list of chemical elements above there has been placed against the name of each substance a number which, for the present, using a term suggested by Davy, we will call its proportional number. Now, the same elementary substances frequently combine with each other in several definite proportions, but these proportions, estimated by weight, are invariably those of these numbers or of their simple multiples. For example, there are two compounds of carbon and oxygen, which contain the relative number of parts, by weight, of each element indicated below:

	Carbon. Parts by weight.	Oxygen. Parts by weight.
Carbonic oxide, ...	12	16
Carbonic dioxide, ...	12	32

There are five compounds of nitrogen and oxygen whose composition in parts, by weight, is as follows:

	Nitrogen. Parts by weight.	Oxygen. Parts by weight.	
Nitrous oxide, . . .	28	16	or 14 : 8
Nitric oxide, ...	14	16	" 14 : 16
Dinitric trioxide, . . .	28	48	" 14 : 24
Nitric dioxide, . . .	14	32	" 14 : 32
Dinitric pentoxide, . .	28	80	" 14 : 40

	Manganese. Parts by weight.	Fluorine. Parts by weight.	
Manganous fluoride, . . .	55	38	$=2\times 19$
Dimanganic hexafluoride, .	55	57	$=3\times 19$
Manganic fluoride, ...	55	76	$=4\times 19$
Dimanganic fluoride, . . .	55	114	$=6\times 19$

Examples like these might be multiplied indefinitely, and the law holds not only when two elements unite, but also when several unite in forming a compound.

There is still another property of these numbers which must not be passed unnoticed, although it is implied in what has already been said. The two numbers, or their multiples, which express the proportions in which each of two elements combines with a third, express also the proportions in which they unite with each other. Thus, 71 parts of chlorine combine with either 32 parts of sulphur or with 56 parts of iron. So, in accordance with the law, 56 parts of iron combine with 32 of sulphur. Again, 14 parts of nitrogen, and also 381 ($=3\times 127$) parts of iodine combine with 3 parts of hydrogen, and so 14 parts of nitrogen unite with 381 of iodine. Lastly, either 16 parts of oxygen, or 32 parts of sulphur, combine with 2 parts of hydrogen, and so 32 parts of sulphur combine with either 32 ($=2\times 16$) parts, or with 48 ($=3\times 16$) parts of oxygen. In the accompanying table these results are given in a tabular form:

32	parts of	sulphur	combine	with	71	parts of	chlorine.
56	“	iron	“	“	71	“	“
56	“	“	“	“	32	“	sulphur.
14	“	nitrogen	“	“	$3 \times 127 = 381$	“	iodine.
3	“	hydrogen	“	“	381	“	“
3	“	“	“	“	14	“	nitrogen.
16	“	oxygen	“	“	2	“	hydrogen.
32	“	sulphur	“	“	2	“	“
32	“	“	“	“	$2 \times 16 = 32$	“	oxygen.
32	“	“	“	“	$3 \times 16 = 48$	“	“

From the facts let us pass, for a moment, to their interpretation, and notice how they at once suggest an atomic theory. To the question which the mind asks, “What mean those definite weights?” the suggestion comes at once, they must mean definite masses of matter; they must be the relative weights of those little masses we have called atoms. And see what a simple interpretation the atomic theory gives of this whole class of phenomena. Assume that there are as many kinds of atoms as there are elementary substances; that all the atoms of the same element have the same weight, and that the “proportional numbers” express the relative weight of the different atoms. Assume further that combination consists merely in the union between atoms, and that chemical changes are determined by their aggregation, separation, or displacement, and we have at once a clear conception of the manner by which the remarkable results we have been studying may be produced. When two elementary substances combine, it must be that a single atom, or some definite number of atoms of one, unite with a definite number of atoms of the other, and therefore the combination must take place either in the proportion of the relative weights of the atoms, or in some simple multiple of that proportion. Moreover, when in any chemical change a new grouping of the atoms takes place, the same relative proportions must be preserved.

From the conception of the atom we naturally return to that of the molecule, in order to discuss the relation between these two quantities, which otherwise we should be liable to confound. You remember the physicist's definition of a molecule: "The small particles of a substance which act as units." The molecules of hydrogen gas are the small, isolated masses of hydrogen, which move like so many worlds through the space occupied by the gas, and, by striking against the walls of the inclosure, produce the pressure which the gas exerts. The molecules of water, in like manner, are the small masses which are driven apart by heat, and become active in the condition of steam. The chemist looks at the molecule from a somewhat different point of view. To him the small masses are not merely centres of forces, but they are the particles in which the qualities of substances inhere. They are the smallest particles of a substance which can exist by themselves. So long as the integrity of the molecule is preserved, the substance is unchanged, but, when the molecules are broken up or changed, new substances are the result. We can carry mechanical division no further than the molecule, but, by chemical means, we can break up the molecules, and the parts of the molecule thus brought to our knowledge are the atoms. Take, for example, common salt:

The smallest particle of this salt which has a salt taste, and in general retains the qualities of salt, is the molecule of salt. This molecule, as we know from the specific gravity of the vapor of salt, weighs 58.5 microcriths. We also know by chemical analysis that, in every 58.5 parts of salt, there are 35.5 parts of chlorine and 23 parts of sodium. Hence, a molecule of salt must contain 35.5 microcriths of chlorine and 23 microcriths of sodium, and, in any chemical process in which chlorine gas or metallic sodium is extracted from salt, each molecule must be subdivided into these two parts. Now, both chlorine gas and sodium are elementary substances, and our theory supposes that the numbers 35.5 and 23 represent the relative weights of their atoms. We, therefore, further conclude that the molecule of salt is formed by the union of two atoms, one of chlorine and one of sodium.

In like manner, the molecules of every compound substance are aggregates of atoms, of at least two atoms each. With the elementary substances it is different. There are many of these whose molecules are never subdivided, and in such cases the molecule and the atom are identical, but there are also several, of which the molecules can be shown to consist of two or more atoms. Thus, the molecules of phosphorus probably consist of four atoms, those of oxygen of two atoms, and those of hydrogen, nitrogen, chlorine, bromine, and iodine, likewise of two.

Assuming that the molecule of hydrogen gas consists of two atoms as just stated, let us dwell on this fact for a moment as explaining our system of estimating molecular weights, which must have appeared, when stated, very arbitrary. You remember that, according to the law of Avogadro, equal volumes of all gases contain, under the same conditions, the same number of molecules. Then, since a given volume of oxygen gas weighs sixteen times as much as the same volume of hydrogen gas, the molecule of oxygen must weigh sixteen times as much as the molecule of hydrogen; and, if we assumed the hydrogen-molecule as our unit of molecular weight, the molecule of oxygen would weigh sixteen of those units. So, also, as nitrogen gas weighs fourteen times as much as hydrogen, the nitrogen-molecule would weigh fourteen of the hydrogen units. Again, as chlorine gas weighs 35.5 times as much as hydrogen, a molecule of chlorine would weigh 35.5 of the same units. But these numbers, 16, 14, and 35.5, are simply the specific gravities of the several gases referred to hydrogen; so that, if we took the hydrogen-molecule as the unit, the specific gravity of a gas or vapor referred to hydrogen would express the molecular weight of the substance in these units. Instead, however, of taking the hydrogen-molecule as our unit, we selected the half-hydrogen molecule for that purpose, and called its weight a microcrith, thus, of course, doubling the numbers expressing the molecular weights. Ten pounds have the same value as twenty half-pounds, and so sixteen hydrogen-molecules have the same value as thirty-two microcriths; and thus it is that, with the system in use, the molecular weight of a substance is twice the specific gravity referred to hydrogen.

Now, you can understand the reason why the half hydrogen-molecule was selected as the unit of molecular weight, and made the microcrith. It was simply because the half-molecule is the hydrogen atom. The microcrith is simply the weight of the hydrogen atom, the smallest mass of matter that has yet been recognized in science. The hydrogen-molecule consists of two atoms, and therefore weighs two microcriths. The oxygen-molecule weighs sixteen times as much as the hydrogen-molecule, and therefore weighs thirty-two microcriths. The specific gravity of carbonic-dioxide gas is 22, that is, it weighs twenty-two times as much as hydrogen. Its molecule is therefore twenty-two times as heavy as the hydrogen-molecule, and, of course, weighs forty-four microcriths. Hence, in general, the specific gravity of a gas referred to hydrogen is the weight of the molecule as compared with the hydrogen-molecule, and twice the specific gravity of a gas referred to hydrogen is the weight of its molecule in hydrogen atoms or microcriths.

But you will ask: How do you know that the hydrogen-molecule consists of two atoms, and, in general, how can you determine the weight of the atom of an element? This is a very important question for our chemical philosophy, and I will endeavor to answer it in the next lecture.

LECTURE VI.

ATOMIC WEIGHTS AND CHEMICAL SYMBOLS.

As I stated in my last lecture, I am to ask your attention at the outset this evening to a discussion of the method by which the chemists have succeeded in fixing what they regard as the weights of the atoms of the several elements. This method is based, in the first place, on the principle that the molecular weight of a substance can be directly inferred from its specific gravity in the state of gas or vapor, the weight of the molecule of any substance in microcriths being equal to twice the specific gravity of the gas or vapor referred to hydrogen. This point has been so fully explained that it is unnecessary for me to enlarge upon it further.

In the second place, our method is based on the principles of what we call quantitative analysis. I have already stated that the chemists have been able to analyze all known substances, and to determine with great accuracy the exact proportions of the several elementary substances which are present in each. The methods by which these results are reached are, for the most part, indirect, and frequently very complicated. They are described at great length in the works on this very important practical branch of our science, but it would be impossible to give a clear idea of them in this connection. It may be well to say, however, that, in order to analyze a substance, it is not necessary actually to extract the several elementary substances and weigh them. Indeed, this can only very rarely be done, but we reach an equally satisfactory result by converting the unknown substance into compounds whose composition has been accurately determined, and from whose weight we can calculate the weights of their elements.

For example, if we wished to determine the amount of sulphur in a metallic ore, we should not attempt to extract the sulphur and weigh it. Indeed, we could not do so with any accuracy; but we should act on a given weight of the ore, say 100 grains, with appropriate agents, and, by successive processes, convert all the sulphur it contained into a white powder called baric sulphate. Now, in accordance with the law of definite proportions, the composition of baric sulphate is invariable, and we know the exact proportion of sulphur it contains. Hence, after weighing the white powder, we can calculate the amount of sulphur in it, all of which, of course, came from the 100 grains of ore.

Evidently, this method assumes an exact knowledge of the amount of sulphur in baric sulphate, which must have been determined previously. This was, in fact, found by converting a weighed amount of sulphur into baric sulphate, and, in a similar way, most of our methods of analysis are based on previous analyses, in which the definite compounds, whose composition we now assume is known, were either resolved into elements or were formed synthetically from the elements.

As the result of such processes as this, we have the relative amounts of the several elements present in the substance analyzed, and it is usual to state the result in per cents. Thus, the analyses of water, salt, and sugar, give the results stated below:

Water.		Salt.		Sugar.	
Hydrogen ...	11.111	Sodium......	39.32	Carbon	42.06
Oxygen	88.889	Chlorine.....	60.68	Hydrogen...	6.50
				Oxygen.....	51.44
	100.000		100.00		100.00

Understanding, then, that we are in possession of means of determining accurately the weights of the molecules of all volatile compounds, and also the exact per cent. of any element which each substance contains, we can readily comprehend the method employed for finding the weight of the atom. Let it be the weight of the oxygen atom which we wish to determine. We compare all the volatile compounds of oxygen as in the diagram (p. 125). We take the specific gravity of their vapors with reference to hydrogen, and, doubling the number thus obtained, we have the molecular weights given in the column under this heading. The analyses of these substances inform us what per cent. of each consists of oxygen. Hence, we know how much of the molecules consists of this element. The weight of oxygen in each molecule is given in the last column, estimated, of course, like the molecular weights, in microcriths. Having thus drawn up our table, let me call your attention to two remarkable facts which it reveals.

Notice, first, that the smallest weight of oxygen in any of these molecules is 16 m.c.; and, secondly, that all the other weights are simple multiples of this.

Here, certainly, is a most wonderful fact. Remember that these numbers, which are displayed here

Atomic Weight of Oxygen.

NAMES OF COMPOUNDS OF OXYGEN.	Weight of molecule.	Weight of oxygen in molecule.
Water	18 m.c.	16 m.c.
Carbonic oxide	28 "	16 "
Nitric oxide	30 "	16 "
Alcohol	46 "	16 "
Ether	74 "	16 "
Carbonic dioxide	44 "	32 "
Nitric dioxide	46 "	32 "
Sulphurous dioxide	64 "	32 "
Acetic acid	60 "	48 "
Sulphuric trioxide	80 "	48 "
Methylic borate	104 "	48 "
Ethylic borate	146 "	48 "
Ethylic silicate	208 "	64 "
Osmic tetroxide	263.2 "	64 "
Oxygen gas	32 "	32 "

Atomic Weight of Chlorine.

NAMES OF COMPOUNDS OF CHLORINE.	Weight of molecule.	Weight of chlorine in molecule.
Hydrochloric acid	36.5 m.c.	35.5 m.c.
Acetylic chloride	78.5 "	35.5 "
Ethylic chloride	64.5 "	35.5 "
Phosgene gas	99. "	71. "
Dicarbonic dichloride	95. "	71. "
Chromic oxychloride	155.2 "	71. "
Arsenious chloride	181.5 "	106.5 "
Boric chloride	117.5 "	106.5 "
Phosphorous chloride	137.5 "	106.5 "
Carbonic tetrachloride	154. "	142. "
Dicarbonic tetrachloride	166. "	142. "
Silicic chloride	170. "	142. "
Tantalic chloride	359.4 "	177.5 "
Columbic chloride	271.4 "	177.5 "
Aluminic chloride	267.8 "	213. "
Dicarbonic hexachloride	237. "	213. "
Chlorine gas	71. "	71. "

so largely, are the results of laborious investigations. Each one of them represents the result of weeks, frequently of months, of labor. The molecular weights were obtained by actually weighing the vapor of each gas, and thus finding its specific gravity; the quantity of oxygen by analyzing each substance, and thus finding the per cent. of oxygen which it contained. Remember that the work has been done at different times, and by many different men, working wholly independently of each other, and with no view to such a result. Now, all this work done, and the results all brought together, it appears that the molecule of every known oxygen compound contains either 16 microcriths of oxygen or some simple multiple of this quantity. It is impossible that this should be a chance coincidence. That invariable repetition of 16 microcriths must have a meaning, and the only explanation we can give is, that it is the weight of definite particles of oxygen, which we call atoms. In other words, then, 16 microcriths, the smallest weight of oxygen known to exist in any molecule, must be the weight of the oxygen atom. In all those molecules, which contain 16 m.c. of oxygen, there is, then, 1 atom of oxygen; in those which contain 32 m.c. of oxygen, there are 2; and, in those which contain 48 m.c., 3 atoms, and so on. Notice also, in this connection, that the molecule of oxygen gas itself weighs 32 m.c., and is, therefore, twice as heavy as the atom. In other words, the molecule of oxygen gas consists of two atoms, and this is one of the cases referred to in the last lecture, in which the molecule of an elementary substance is not the same as the atom.

Take, now, another elementary substance—chlorine. Here we have a list of some of the volatile compounds of this element. As before, the molecular weights annexed were determined by doubling the known specific gravities of the vapors of the several substances, and the weight of chlorine in each molecule was calculated from the results of oft-repeated analyses. Notice that the smallest weight of chlorine in a molecule is 35.5 microcriths, and that the other molecules have either the same weight or a simple multiple of it. This number, 35.5, appears here with the same constancy as the number 16 in the previous table. As before, this constancy cannot be an accident. These 35.5 microcriths of chlorine must be definite masses of the elementary substance, which retain their integrity under all conditions, and are not subdivided in any known chemical changes, and these wonderfully minute but definite masses are what we call the chlorine atoms. The atoms of chlorine, therefore, weigh 35.5 microcriths. Hence, the molecule of hydrochloric acid contains one chlorine atom, the molecule of phosgene gas two such atoms, the molecule of boric chloride three, that of silicic chloride four, and that of aluminic chloride six. Lastly, as in the case of oxygen, the molecule of chlorine gas is twice as heavy as the atom, or, as we say, consists of two atoms.

Consider, now, the facts in regard to volatile compounds of carbon as they are shown in the next diagram. Here we have a similar constancy in the repetition of the number 12. Twelve microcriths is the smallest quantity of carbon contained in the molecule of any compound of this element whose molecular weight has been determined; and all molecules of carbon compounds, whose weight is known, contain either 12 microcriths of the elementary substance, or else some whole multiple of 12 microcriths. Again the question forces itself upon us, What means this won-

Atomic Weight of Carbon.

Names of Compounds of Carbon.	Weight of Molecule.	Weight of Carbon in Molecule.
Marsh-gas........................	16 m.c.	12 m.c.
Olefiant gas	28 "	24 "
Propylic alcohol	60 "	36 "
Ether............................	74 "	48 "
Amylic alcohol....................	88 "	60 "
Triethylstibine....................	209 "	72 "
Toluol...........................	98 "	84 "
Oil of wintergreen	152 "	96 "
Cumol...........................	120 "	108 "
Oil of turpentine..................	136 "	120 "
Amyl benzol	148 "	132 "
Diphenylamine	169 "	144 "

Atomic Weight of Hydrogen.

Names of Compounds of Hydrogen.	Weight of Molecule.	Weight of Hydrogen in Molecule.
Hydrochloric acid	36.5 m.c.	1 m. c.
Hydrobromic acid	81. "	1 "
Hydriodic acid	128. "	1 "
Hydrocyanic acid	27. "	1 "
Water...........................	18. "	2 "
Hydric sulphide..................	34. "	2 "
Hydric selenide..................	81.5 "	2 "
Formic acid......................	46. "	2 "
Ammonia gas....................	17. "	3 "
Hydric phosphide................	34. "	3 "
Hydric arsenide	78. "	3 "
Acetic acid......................	60. "	4 "
Olefiant gas.....................	28. "	4 "
Marsh-gas	16. "	4 "
Alcohol.........................	46. "	6 "
Ether	74. "	10 "
Hydrogen gas....................	2. "	2 "

derful constancy? Does any one suspect that it may be a fiction of our scientific theorizing—a mere play with numbers? Let him only acquaint himself with the facts, and he will find how groundless his suspicion is. The evidence of these facts is far stronger than would appear from our table. The number of volatile carbon compounds is very large, and our list might have been greatly extended. It must also be constantly remembered, as I have said, that these tables embody the result of a vast amount of experimental labor—labor, I may add, without price, and whose only object was the truth. Now, all this labor done, these wonderful results appear. We must explain them; and the only explanation we can give is, that the molecules of these carbon compounds are formed of small masses of the elementary substance which weigh twelve microcriths, and these small masses are the carbon atoms.

Before leaving the subject, let me call your attention to one other table, in which similar facts in regard to the volatile compounds of hydrogen have been collated. Like the last, this table might have been greatly extended; but a sufficient number of facts have been collected to show that the smallest quantity of hydrogen, in any molecule, weighs one microcrith, and that the quantities of this elementary substance in the molecules of its various compounds are in all cases whole multiples of this small mass, which we call the hydrogen atom. The hydrogen atom, then, weighs one microcrith, and the several molecules contain as many hydrogen atoms as they contain microcriths of hydrogen. Hence, the molecule of hydrogen gas, which weighs two microcriths, consists of two atoms. The hydrogen atom is the smallest mass of matter known to science, and I hope you can now appreciate the reason why it has been chosen as the unit of molecular and atomic weights. I also hope that I have been able to convince you that it is a definite mass of matter, and that we have as much right to name it a microcrith as to call a certain mass of metal a grain, or another mass a pound.

In a similar way the weights of the atoms of several of the other elementary substances have been determined; but the method is not universally applicable, for there are many of the elementary substances which do not yield a sufficient number of volatile bodies to enable us to fix the molecular weight of as many of their compounds as would be required to make our conclusion trustworthy. In such cases, however, we have other methods of finding the molecular weight, which, although not so fundamental or so simple as that based on the specific gravity of the vapor, give for the most part satisfactory results. These methods, however, would not be intelligible at the present stage of our study.[1]

I trust we are all now prepared to understand the significance of the numbers, which, in the table of chemical elements (on page 112), are associated with the names of the elementary substances.

[1] The molecular weights given in the tables on pages 125 and 128 are not in most cases the exact values, which would be obtained by doubling the specific gravities actually found by experiment, but they are those values corrected by the methods alluded to above. The subject is complex, involving the relative value and degree of accuracy of two kinds of experimental evidence, and its premature discussion at this time would only serve to confuse the reader. It is sufficient for the present to say, that the correction here referred to does not in the least degree invalidate the conclusions we have drawn from the tables, and this will be seen to be the case when the subject is fully understood.

These numbers represent the weights of the several atoms in microcriths.

As I have said, the idea that the atoms are isolated masses of matter may be a delusion, and so, as I have also intimated, we may doubt whether the magnitudes in optics, known as wave-lengths, are the lengths of actual ether-waves; but, just as these magnitudes are definite values, on which we can base calculations with perfect confidence, although the form of the magnitude may not be known, so the atomic weights are invariable quantities, whose relative values are as well established as any data of science; and, however our theories in regard to them may change, they must always remain the fundamental constants of chemistry. On these data are based all those calculations by which we predict the quantitative relations of chemical phenomena, and, starting from the new stand-point which they furnish, we shall now proceed to develop still further the philosophy of our science.

But, before we go forward, let me call your attention to a very striking coincidence, which greatly tends to confirm the general correctness of the results we have reached:

You are well aware that the amount of heat required to raise the temperature of the same weight of material to the same degree differs very greatly for different substances. In order to secure a standard of reference, it has been agreed to adopt, as the *unit of heat*, the amount of heat-energy required to raise the temperature of one pound of water one Fahrenheit degree, or, in the French system, one kilogramme of water one centigrade degree. As water has a greater capacity for heat than any substance known (except hydrogen gas), it requires only a fraction of a unit of

Specific Heat of Elementary Substances.

	Specific Heat.	Atomic Weight.	Products.
Lithium	0.9408	7.	6.59
Sodium	0.2934	23.	6.75
Magnesium	0.2499	24.	6.00
Aluminum	0.2143	27.5	5.89
Phosphorus	0.1887	31.	5.85
Sulphur (native)	0.1776	32.	5.68
Potassium	0.1696	39.	6.61
Manganese	0.1217	55.	6.69
Iron	0.1138	56.	6.37
Nickel	0.1108	58.7	6.50
Cobalt	0.1073	58.7	6.30
Copper	0.0951	63.5	6.04
Zinc	0.0955	65.2	6.26
Arsenic	0.0814	75.	6.11
Selenium (metallic)	0.0761	79.4	6.02
Bromine (solid)	0.0843	80.	6.75
Molybdenum (impure)	0.0722	96.	6.93
Rhodium	0.0580	104.4	6.07
Palladium	0.0593	106.6	6.32
Silver	0.0570	108.	6.16
Cadmium (impure)	0.0567	112.	6.35
Tin	0.0562	118.	6.63
Antimony	0.0508	120.3	6.11
Iodine	0.0541	127.	6.87
Tellurium	0.0474	128.	6.06
Tungsten	0.0334	184.	6.15
Gold	0.0324	197.	6.38
Platinum	0.0324	197.4	6.39
Iridium	0.0326	198.	6.45
Osmium	0.0311	199.2	6.20
Mercury (solid)	0.0319	200.	6.38
Thallium	0.0335	204.	6.84
Lead	0.0314	207.	6.50
Bismuth	0.0308	210.	6.48
Boron (crystallized)	0.2500	11.	2.75
Carbon (diamond)	0.1469	12.	1.76
Carbon (graphite)	0.2008	12.	2.41
Carbon (wood charcoal)	0.2415	12.	2.90
Silicon (crystallized)	0.1774	28.	4.97

heat to raise the temperature of one pound of any other substance one degree. This fraction is called the *specific heat* of the substance, and its value has been determined experimentally, with great care, for a very large number of substances, including most of the elementary substances. In the second column in the table on the opposite page we have given the specific heat of more than one-half of the elementary substances. We owe these results to Regnault, and his investigations on this subject are among the most important of the many valuable contributions to science of this eminent French physicist. As the specific heat of a substance in different states of aggregation often varies very greatly, only the results obtained with the elementary substances in the solid state are here given, and the numbers in each case stand for the fraction of a unit of heat required to raise the temperature of one pound of the solid one degree. The figures in the second column of our table are the atomic weights of the elements, and those in the third column the products obtained by multiplying these weights by the specific heat. Notice how constant this product is. It varies only between 5.7 and 6.9, and there are strong reasons for believing that the variations depend on differences in the physical condition of the elementary substances. We know that this condition very greatly influences the thermal relations of solid bodies, and, if the substances could be compared in precisely the same state, it is possible that the above product would be found to be absolutely constant, the most probable value being 6.34. Only three solid elementary substances are known the product of whose atomic weight by the specific heat does not fall within the limits assigned above, and these are the different forms of carbon, boron, and silicon, all elements remarkable for the wide differences between the physical conditions in which they are known.

What, now, can be the explanation of the remarkable law which the table presents to our notice? The usual explanation is, that the atoms of the different elements have the same capacity for heat, and hence, that masses of the elementary substances containing the same number of atoms must have the same capacity for heat when under similar physical conditions; the constant product being the amount of heat required to raise the temperature of such masses to the same degree. If, for example, it requires the same amount of heat to increase by one degree the temperature of either 56 m.c. of iron (one atom) or 200 m.c. of mercury (also one atom), it will also require equal amounts to raise the temperature of 56 pounds of iron and 200 pounds of mercury one degree; and hence 56×0.1138 (the specific heat of iron) must be equal to 200×0.0319 (the specific heat of mercury).—You will remember, of course, that the decimal in each case represents the fraction of a unit of heat required to raise the temperature of *one pound* one degree.

But, all theorizing apart, an agreement like this cannot be the result of accident; and, even if we cannot explain the law, the very coincidence gives us great confidence in the values of the atomic weights we have adopted.

Let us now, for a moment, recapitulate. All substances are collections of molecules, and in these molecules their qualities inhere. What is true of the substance is true of the molecule. The molecule is an aggregate of atoms; sometimes of atoms of the same kind, as in elementary substances, sometimes of atoms of different kinds, as in compound substances. The molecules are destructible, while the atoms are indestructible; and chemical change consists in the production of new molecules by the rearrangement of the atoms of former ones. Such, then, are our conceptions of the constitution of substances, and I next proceed to show how we are able to represent this constitution by means of a most beautiful system of notation, with which you must be all more or less familiar, under the name of chemical symbols.

Just as in algebra letters are used to represent quantities, so in chemistry we use the initial letters of the Latin name of the elementary substance to represent that mass of each element we call an atom. Thus, O represents one atom of oxygen, N one atom of nitrogen, C one atom of carbon, Cl one atom of chlorine, Cr one atom of chromium, F one atom of Fluorine, Fe one atom of ferrum (iron), S one atom of sulphur, Sb one atom of stibium (antimony). By using the first letters of the Latin names, a uniformity has been secured among all nations, the convenience of which is obvious, and it is only in a few cases that the Latin initial differs from the English. These symbols necessarily represent a definite weight, that is, the weight of the atom. O stands for 16 microcriths of oxygen, C for 12 microcriths of carbon; and, in each case, the symbol stands for the atomic weight given in our table (page 112). In order to represent several atoms, we use figures placed, like algebraic exponents, above or below the symbol. These exponents do not, as in algebra, indicate powers, but only multiples; thus, O_2 means two atoms, or 32 m.c. of oxygen, C_6 six atoms, or 72 m.c. of carbon, and so on.

Having adopted this simple notation for the atom,

we easily represent a molecule by writing together the symbols of the atoms of which it consists, indicating the number of each kind of atoms by figures, as above. A molecule of water, for example, consists of three atoms, two of hydrogen and one of oxygen. Hence, its symbol is H_2O. This symbol shows, not only that the molecule consists of three atoms, as just stated, but also that it contains 2 m.c. of hydrogen and 16 m.c. of oxygen. Further, it shows that the molecule of water weighs 18 m.c. If we wish to represent several molecules of water, we place a figure before the whole symbol. Thus, $2H_2O$ represents two molecules of water, $5H_2O$ five molecules of water, etc. Now, since, in all chemical relations, what is true of the molecule is true of the substance, this symbol may be regarded as the symbol of water, and is constantly spoken of as such. Again, a molecule of alcohol is known to consist of two atoms of carbon, six atoms of hydrogen, and one of oxygen. Hence, the symbol of the molecule is C_2H_6O. This symbol informs the chemist that a molecule of alcohol contains 2 atoms or 24 m.c. of carbon, 6 atoms or 6 m.c. of hydrogen, and 1 atom or 16 m.c. of oxygen. It also shows that the total weight of the molecule is 46 m.c. Several molecules of alcohol are indicated by the use of coefficients, as before—thus $3C_2H_6O$, etc. This is the whole of the system, and you see how beautiful and simple it is. The single letters stand for atoms, and the terms formed by the grouping of the letters stand for molecules, and the very possibility of the system is in itself a very strong proof that molecules and atoms really exist.

Before proceeding to show how admirably this system is suited to express chemical changes, let me ask your attention for a moment to the nature of the evidence by which the symbol of a substance is fixed; for, although this evidence is precisely of the same kind as that on which the atomic weights of the elementary substances rest, yet the principles involved are so important that a brief restatement of the evidence, as it bears on the present problem, seems almost necessary for a clear understanding of the subject. The question is this: What is your proof that the symbol of alcohol, for example, is C_2H_6O, or, in other words, that this symbol represents the constitution of a molecule of alcohol? The evidence is—

1. We know by experiment (page 79) that the specific gravity of alcohol-vapor referred to hydrogen is 23. Hence, since, by Avogadro's law alcohol-vapor and hydrogen gas have in the same volume the same number of molecules, the molecule of alcohol is twenty-three times as heavy as the molecule of hydrogen gas; and, further, since by assumption the hydrogen-molecule weighs 2 m.c., the alcohol-molecule weighs 46 m.c.

2. We have analyzed alcohol, and know that it has the following composition:

Analysis of Alcohol.

	Per cent.	Composition of molecule.
Carbon........................	52.18	24 m.c.
Hydrogen........................	13.04	6 "
Oxygen........................	34.78	16 "
	100.00	46 "

Hence, of the molecule of alcohol $52\frac{18}{100}$ per cent., or 24 parts in 46, consist of carbon, $13\frac{4}{100}$ per cent., or 6 parts in 46, consist of hydrogen, and $34\frac{78}{100}$, or 16 parts in 46, consist of oxygen. The whole adds up, as you see, 46, showing that we have done our sum correctly.

Analysis, then, proves that, of the molecule of alcohol weighing 46 m.c., 24 m.c. are carbon, 6 m.c. are hydrogen, and 16 m.c. are oxygen. But the weight of an atom of carbon is 12 m.c., hence the molecule contains two atoms of carbon, or C_2; the weight of an atom of hydrogen is 1 m.c., hence the molecule contains 6 atoms of hydrogen, or H_6; the weight of the oxygen atom is 16 m.c., hence the molecule contains one atom of oxygen, or O, and the symbol is C_2H_6O.

Again, why is the symbol of water H_2O? 1. The specific gravity of steam referred to hydrogen gas is 9, hence the weight of a molecule of water in microcriths is 18. 2. Analysis shows that water has the following composition in 100 parts:

Analysis of Water.

	Per cent.	Composition of molecule.
Hydrogen	11.11	2 m.c.
Oxygen	88.89	16 "
	100.00	18 "

We know, then, that, of the molecule weighing 18 m.c. of water, $11\frac{11}{100}$ per cent., or 2 m.c., consist of hydrogen, and $88\frac{89}{100}$ per cent., or 16 m.c., consist of oxygen. But 2 m.c. of hydrogen equal 2 atoms, or H_2, and 16 m.c. of oxygen 1 atom, or O. Hence, the symbol is H_2O.

You see how simple is the reasoning and how definite the result; and, unless our whole theory in regard to molecules and atoms is in error, there is no more doubt that the symbol of water should be written H_2O, than that this familiar liquid consists of oxygen and hydrogen gas.

But many of my audience will remember that, when they studied chemistry, the symbol of water was

HO, and will ask, Why this change? I answer: This difference is of a type with the whole difference between the old and the new schools of chemistry. Indeed, the two symbols may be regarded as the shibboleths of the two systems. In the old system, the symbols simply stood for proportions, and nothing else. The symbol H meant 1 part by weight of hydrogen, and O 8 parts by weight of oxygen; and HO meant a compound, in which the two elements were combined in the proportions of 1 to 8, which is as true of water now as it was then. In the old system, the special form of the symbol, whether H_2O, HO, or HO_2, had no significance, for this was determined by the arbitrary values given to the letters. There is a second compound of hydrogen and oxygen called hydric peroxide, in which the elements are combined in the proportion of 1 of hydrogen to 16 of oxygen; and, had the chemists of the old school assigned to the symbol O the value 16 instead of 8, then the symbol of hydric peroxide would have been written HO, and that of water H_2O; and the only reason usually given for making O represent 8 parts of oxygen instead of 16 was, that water, being very widely diffused in Nature, and the most stable compound of the two, ought to be represented by the simplest symbol; or, in other words, that the ratio between the quantities of oxygen and hydrogen, which it contains, ought to be taken as the type ratio between these elements.

This reasoning was as unsatisfactory as it has proved to be unsound. It might justly have been said that the system, although artificial, was consistent in itself, and that it better suited the requirements of the system to assign to oxygen the proportional number 8, than to select a multiple of that number. Indeed, this was the light in which the whole scale of proportional numbers was regarded by a large majority of the students of chemistry during the first half of this century; and it is only necessary to state that the German chemists, following the lead of Berzelius, used for years a scale in which oxygen was taken as 100, in order to show how purely arbitrary the actual numbers were considered to be. The only truth that the numbers were believed to represent was the law of definite and multiple proportion; and, so long as the true proportions were preserved, any scale of numbers might be used which suited the experimenter's fancy.

It is, however, perfectly true that, in selecting one of several multiples, which might be used for a given element in a given scale, the decision of the chemist was not unfrequently influenced by the very ideas which now form the basis of our modern science; as is shown by the fact that the proportional numbers of Davy and Berzelius were called chemical equivalents by Wollaston, and atomic weights by Dalton and his pupils. But, then, the truths, which these terms now imply, were never fully conceived or consistently carried out. The atomic weights of the new system are the weights of real quantities of matter, the combining numbers of the old system were certain empirical proportions. So is it in other particulars, and the difference between the new school and the old is really the difference between clear and misty conceptions.

Our modern science is a philosophical system, based on ideas distinctly stated and consistently developed. The chemists of the old school can hardly be said to have had a philosophy, but they had an admirable nomenclature, which was almost as good as a philosophy, and served to classify the facts while the fundamental principles of the science were being slowly developed. It was, of course, to be expected that the fundamental ideas of our science should be conceived separately and at first only imperfectly; and it was not until clear and definite conceptions had been reached, and the relations of the several ideas clearly understood, that a philosophy of chemistry was possible. Of course, we are far from believing that the ideas, now prevailing, are necessarily true, and it is perhaps to be expected that our modern school will share the same fate as that which preceded it; but we do believe that the coming system, whatever it may be, will be based on equally clear conceptions, and that, in attempting to clarify our ideas and realize our conceptions, we are following the right path, and making the only satisfactory progress.

Before closing the lecture, it only remains for me to show how the system of notation I have described may be used to express chemical changes, and I can best illustrate this use by applying it in a practical example. The experiment I have selected for the purpose must be familiar to every one in some form or other.

In the first place, we have in this large glass vessel a white, pulverulent solid, familiarly called soda. The chemists call it sodic carbonate. It consists of molecules, which are each formed of six atoms, two of a metal called sodium, one of carbon, and three of oxygen. Hence, the symbol is Na_2CO_3. In the second place, we have in this pitcher a liquid well known in commerce under the name of muriatic acid. It is a solution in water of a compound which is called in chemistry hydrochloric acid. Hydrochloric acid itself, as I shall show you at the next lecture, is a gas $18\frac{1}{4}$ times as heavy as hydrogen; hence its molecular weight is $36\frac{1}{2}$—and its molecules, as is well known, consist of one atom of chlorine and one of hydrogen. Its symbol is then HCl—and the condition of aqueous solution we may express by the addition of the letters Aq, the initial of *aqua*, the Latin name of water—thus: HCl+Aq.

On pouring the acid upon the soda, there is at once a violent effervescence; and a large quantity of gas is evolved, which will soon fill the glass jar. The old substances disappear, and new substances are formed. This, then, is a chemical change. Such a change, in the language of chemistry, is called a reaction. The substances taking part in the change are called the factors, and the substances formed are called the products of the reaction.

In the present example, the factors are sodic carbonate, hydrochloric acid, and water. What are the products?

First of all, we have a large volume of colorless gas, and not only a large volume, but also a very considerable weight, since, for a gas, it is quite a heavy substance. In old times this product of the process was wholly overlooked; but I can easily prove to you that there is a no inconsiderable amount of material in the upper part of this glass vessel, although in an invisible condition. First, by lowering a lighted candle into the jar, I can show that the air has been displaced by a medium in which the candle will not burn. In the second place, by dipping out some of the gas and pouring it into this paper bucket, I can make evident that its weight is appreciable: You notice that the end of the balance-beam to which the bucket is suspended immediately falls; and see, also, how these candles are extinguished, as the heavy gas from my dipper flows down on the flames. Lastly, by repeating the experiment on a smaller scale in front of the lantern, and projecting the image of the small glass vessel, we here use, on the screen, I can make the current of gas visible as it flows over the lip.

This aëriform material is now called in chemistry carbonic dioxide, but you are more familiar with it under the old name of carbonic acid. It is the chief product of the burning of coal and wood; and, when you are told that every ton of coal burned yields $3\frac{2}{3}$ tons of this gas, you can conceive what immense floods are being constantly poured into the atmosphere from the throats of our chimneys. It is also being continually formed, and in still greater amounts, by the processes of respiration, fermentation, and decay. Although familiarly known only in the state of gas, it can readily be reduced by pressure and cold to the liquid condition; and, when in this condition, is easily frozen, forming a transparent solid like ice, or a loose, flocculent material like snow, under different conditions. It is a compound simply of carbon and oxygen, and no fact of chemistry is better established than that every molecule of this gas consists of one atom of carbon and two atoms of oxygen. Hence its symbol is CO_2.

The presence of the other products formed in our experiment I cannot make so readily evident to you, although they are really far more tangible than this gas. One of them is water, which at once mingles with the large body of water used in the experiment. The other is common salt. This dissolves, as it forms, in the water present; but, after the reaction is ended, it can easily be isolated by evaporating the brine. We will start the process, so that any one who is skeptical can satisfy himself, by tasting the residue, that common salt has been really formed.

Common salt is composed of a metal, sodium, and chlorine gas. Its molecules are known to consist, each of an atom of sodium and an atom of chlorine. Hence its symbol is NaCl.

Let us now write the factors of this reaction opposite to the products, so that we can compare them :

Na_2CO_3	HCl	$NaCl$	H_2O	CO_2.
Sodic Carbonate.	Hydrochloric Acid.	Sodic Chloride, or Common Salt.	Water.	Carbonic Dioxide.

Now, let me remind you of a simple principle, which we must apply in interpreting this reaction. No material can be lost. These atoms are indestructible, so far as we know. If, then, we have here all the factors and all the products (and there can be no doubt whatever on this point), there must be just as many atoms of each element in the products as there are in the factors, and *vice versa.* Now, there are two atoms of sodium in the molecule of sodic carbonate. Hence there must be two atoms of the same element in the products, and we must therefore write 2NaCl. The molecule of water in the products has two atoms of hydrogen ; hence we must write 2HCl among the factors. Thus amended, our reaction becomes :

$$Na_2CO_3 + 2HCl = 2NaCl + H_2O + CO_2.$$

Now, since the quantity of material represented among the products exactly equals that represented among the factors, we may very properly employ the equation-sign of algebra to separate the two members of our reaction ; and, further, it becomes equally natural to separate the several terms by the plus-sign. When, now, we study the chemical change, as thus written out for our inspection, we see that, in the process, each molecule of sodic carbonate is acted upon by two molecules of hydrochloric acid. The two atoms of sodium (Na_2) from the molecule of sodic carbonate (Na_2CO_3) unite each with an atom of chlorine (Cl) from the two molecules of hydrochloric acid ($2HCl$), and there are thus formed two molecules of common salt ($2NaCl$). Meanwhile, the original molecules having been broken up, the other atoms group themselves together to form a molecule of water (H_2O) and a molecule of carbonic dioxide (CO_2). In a word, the chemical change consists in the breaking up of the old molecules and the rearrangement of the atoms to form others, and you will notice how perfectly our system of symbols enables us to follow the steps of the process.

In saying that this equation represents the process, we assume the truth of the principle, already so often reiterated, that what is true of the molecules is true of the substances. Our equation merely represents the reaction between one molecule of sodic carbonate and two of hydrochloric acid. Of course, there were billions on billions of molecules in our glass jar, but then the action here represented was simply so many billion of times repeated.

There is only one other point in connection with this experiment to which I wish to call your special attention before closing the lecture. We used a great deal of water in the process, and the experiment would not have succeeded without it. Now, what part does the water play? An essential part—and this point has a most important bearing on our theory of molecules.

The reaction we have been studying takes place, as we have said, between molecules. But, in order that the molecules of the one body should act on those of the other, it is obviously necessary that they should have a certain freedom of motion. If the molecules had been rigidly fixed in the material of the two substances, it would obviously have been impossible for them to marshal themselves in the manner we have described, two of one substance associating with one of the other in the resulting chemical process. Now, in a solid body, the molecules are to a great extent fixed, and hence no chemical action is possible between such substances, except to a limited extent. There are, in general, two ways by which the required freedom of motion can be obtained: One is to convert the substance into vapor, when, as we have seen, the molecules become completely isolated, and move with great velocity through space, their motion being only limited by the walls of the containing vessel; but this method is only applicable to volatile bodies. The second method is to dissolve the solid in some solvent, when the molecules, as before, become isolated, and move freely through the mass of the liquid. The last is the method generally used, and water, being such a universal solvent, is the common vehicle employed to bring substances together, and for that reason it enters into a very great number of chemical changes. Such was its office in the process we have been studying. We dissolved both the sodic carbonate and the hydrochloric acid in water, in order that their molecules might readily coalesce. An experiment will enforce the principle I have been enunciating:

There are a great many substances which will act on sodic carbonate like hydrochloric acid; for example, almost all the so-called acids or acid salts, and, among others, that white solid with which you are familiar under the name of cream-of-tartar. Here we have cream-of-tartar and sodic carbonate, both in fine powder, and we have been carefully mixing them together in this mortar. You see, there is no action whatever; and, in a dry place, we can keep the mixture indefinitely without change. If, however (placing the mixture in this glass vessel), we pour water over it, we have at once a brisk effervescence, and carbonic dioxide is evolved as before. It required the water to bring the molecules together.

Since, then, the water plays such an important part in the reaction, I prefer to indicate its presence, and this may be done by using the symbol Aq. as previously described.

$$\underset{\text{Solution of Sodic Carbonate and Hydrochloric Acid.}}{(Na_2CO_3 + 2HCl + Aq.)} = \underset{\text{Solution of Common Salt.}}{(2NaCl + H_2O + Aq.)} + \overline{CO_2}.$$

This indicates not only that both of the factors are in solution, but also that we have, as one of the products, a solution of common salt. That the second product, carbonic dioxide, is a gas, I sometimes indicate by a line drawn over the symbol, as above.

The second reaction is equally simple, but cream-of-tartar has a vastly more complex molecule than HCl. Its symbol is $HKC_4H_4O_6$, that is, each molecule consists of four atoms of carbon, six atoms of oxygen, one atom of potassium, and five atoms of hydrogen. I write one of the atoms of hydrogen apart from the rest, because it has a very different relation to the molecule—a relation which I shall hereafter explain. The reaction would be written thus:

$$(Na_2CO_3 + 2HKC_4H_4O_6 + Aq.) =$$
$$\underset{\text{Solution of Rochelle Salts.}}{(2NaKC_4H_4O_6 + H_2O + Aq.)} + \overline{CO_2}.$$

With this reaction many of my audience must be familiar, as a mode of raising dough in the process of making bread. The first member of the equation indicates that the two substances are used in solution. There is formed, as the product of the reaction, besides the carbonic dioxide gas, which puffs up the dough, the solution of a salt, whose molecule has the complex constitution I have indicated, and which is a well-known medicine under the name of Rochelle-salts. When soda and cream-of-tartar are used in making bread, this salt remains in the loaf. The amount formed is too small to be injurious, but I cannot but think, although it may be a prejudice, that chemicals had better be kept out of the kitchen.

nite weights of the quantities they represent. Each symbol stands for the known weights of the atoms which are tabulated in this diagram (table, page 112), and the weights of the molecules, which the several terms represent, are found by simply adding up the weights of the several atoms of which they consist. When the substance is capable of existing in the aëriform condition, its molecular weight can be found, as I have shown, from its specific gravity; but these symbols assume that either by this or by some other method the constitution of the molecule has been determined; and, now that the result is expressed in symbols, nothing is easier than to interpret what they have to tell us. To begin with the sodic carbonate, Na_2CO_3. The weight of this molecule is $2 \times 23 + 12 + 3 \times 16 = 46 + 12 + 48 = 106$ m.c. The weight of the molecule HCl is $1 + 35.5 = 36.5$, and two such molecules would weigh 73 m.c. Next, for the products, we have $NaCl = 23 + 35.5 = 58.5$, and $2NaCl = 117.0$, also $CO_2 = 12 + 32 = 44$, and $H_2O = 2 + 16 = 18$. Hence the terms of our equation stand for the weights written over them below:

$$(\overset{106}{Na_2CO_3} + \overset{73}{2HCl} + Aq.) = (\overset{117}{2NaCl} + \overset{18}{H_2O} + Aq.) + \overset{44}{CO_2}.$$

We leave out of the account the water represented by Aq., for this, being merely the medium of the reaction, is not changed. Now we can prove our work; because, if we have added correctly, the sum of the weights of the factors must exactly equal the sum of the weights of the products—and so it is $106 + 73 = 179$, and $117 + 18 + 44 = 179$. Besides the information which the equation gives us in regard to the manner in which the chemical change takes place, the symbols also inform us that 106 parts by weight of sodic carbonate are acted upon by 73 parts by weight of hydro-

LECTURE VII.

CHEMICAL REACTIONS.

To master the symbolical language of chemistry, so as to understand fully what it expresses, is a great step toward mastering the science; and so important is this part of my subject that I propose to occupy the hour this evening with a number of illustrations of the use of symbols for expressing chemical changes.

First, I will recur to the experiment of the last lecture, for we have not yet learned all that it is calculated to teach.

Let us again write on the black-board the symbols which represent the chemical process:

$$\underset{\text{Sodic Carbonate.}}{(Na_2CO_3} + \underset{\text{Hydrochloric Acid.}}{2HCl} + Aq.) = (\underset{\text{Common Salt.}}{2NaCl} + \underset{\text{Water.}}{H_2O} + Aq.) + \underset{\text{Carbonic Dioxide Gas.}}{\overline{CO_2}}.$$

We bring together a solution of sodic carbonate and hydrochloric acid; and there are formed as products a solution of common salt, water, and carbonic dioxide gas. I need not refer again to the circumstance that the state of solution is an essential condition of the change, for this point was fully discussed at the time; but, before we pass on to another experiment, I wish to call your attention to the fact that the several terms in this equation stand for absolutely defi-

chloric acid, and that the yield is 117 parts of common salt, 18 parts of water, and 44 parts of carbonic-dioxide gas.

We learn from this, in the first place, the exact proportion in which the sodic carbonate and hydrochloric acid can be most economically used; for, if the least excess of one or the other substance over the proportions indicated is taken, that excess will be wasted. It will not enter into the chemical change, but will be left behind with the salt and water.

Assume, then, that we have 500 grammes of sodic carbonate, and we wish to know what amount of hydrochloric acid to use, we simply make the proportion as $106 : 73 = 500 : x = 344\frac{36}{106}$. Again, suppose we wish to know how much common salt would be produced from these amounts of sodic carbonate and acid, we write a similar proportion—

$$106 : 117 = 500 : x = 552, \text{ nearly.}$$

So, then, in any process, after we have written the reaction as above, if the weight of any factor or product is given, we can calculate the weight of any other factor or product by this simple rule:

As the total molecular weight of the substance given is to the total molecular weight of the substance required, so is the given weight to the required weight. By total molecular weight we mean, evidently, not the weight of a single molecule, but the weight of the number of molecules which the equation indicates.

This may be called the golden rule of chemistry.

In the laboratory we never mix our materials at random, but always weigh out the exact proportions found by this rule. When one of the products is a gas, as in the present case, a simple modification of the rule enables us to calculate the volume of the resulting gas. Suppose, for example, we wished to calculate what volume of carbonic-dioxide gas could be obtained from 500 grammes of sodic carbonate. We should first find the weight by the above rule:

$$106 : 44 = 500 : x = 207\tfrac{1}{2}, \text{ nearly.}$$

The answer is 207½ grammes of carbonic dioxide. To find the corresponding volume in litres, we have merely to divide this value by the weight of one litre of the gas. Now, there are tables, in which the weight of one litre of each of the common gases is given; but such tables, although convenient, are not necessary, when, as in a written reaction, we know the molecular weights of the substances with which we are dealing. You remember that the molecular weight is always twice the specific gravity with reference to hydrogen. Half the molecular weight is, then, the specific gravity with reference to hydrogen. For example, the molecular weight of carbonic dioxide (CO_2) is 44, and its specific gravity with reference to hydrogen 22—in other words, a litre of carbonic dioxide weighs 22 times as much as a litre of hydrogen. Now, a litre of hydrogen, under the normal pressure of the atmosphere, and at the freezing-point of water, weighs one crith, or 0.0896 gramme, or, near enough for common purposes, 0.09 gramme. If, then, a litre of carbonic dioxide is 22 times as heavy, its weight is 22 criths, or $22 \times 0.09 = 1.98$ gramme. Our total product, above, being 207½ grammes, the number of litres will be $207\tfrac{1}{2} \div 1.98$, or very nearly 104 litres. A litre, as I have said, is very nearly 1¾ pint, but we always use these French weights and measures in the laboratory, so that the values are as significant to the chemist as are pounds and pints to the trader. The general rule, then, is this: We first find the weight of one litre of the gas in grammes, by simply multiplying one-half of its molecular weight by $\frac{9}{100}$, and then we reduce the weight of the gas in grammes to litres by dividing the weight by this product.

Let us pass, now, to another case of chemical change, and the example which I have selected is closely related to the last. One of the products of that reaction was carbonic-dioxide gas, and here we have a jar of that aëriform substance. On the other hand, I have in this bottle an elementary substance, called sodium. It belongs to the class of metals, and is one of the constituents of sodic carbonate, which we used in the former experiment. I now propose to cause these two substances to act chemically upon each other; but, as before, no chemical action will result unless the molecules have sufficient freedom of motion. Those of the carbonic-dioxide gas are already as free as the wind, moving with immense velocity through this jar. But not so with those of the sodium. In the usual solid condition of this metal, the motion of its molecules is restricted within very narrow limits. Before, we gave freedom to the molecules of sodic carbonate and hydrochloric acid by dissolving the substances in water. That method is not applicable here, for sodium acts chemically on water, and with great violence; but we can reach a similar result by melting the sodium, and heating the molten metal until it begins to volatilize. Then, on introducing the crucible containing the seething metal into the gas, the molecules of the sodium, as they are forced up by the heat, will come into contact with those of the carbonic dioxide, and a violent chemical action will be the result.

This action is made evident to you by the brilliant light evolved, and the sodium, as you would say, is burning in the carbonic-dioxide gas. Let us now represent this chemical change by our symbols.

Beginning with the factors, the molecule of carbonic dioxide, as already stated, is represented by the symbol CO_2. The weight of the molecule of sodium has not yet been accurately determined; and, in the absence of exact information, we will assume, as is most probable, that the molecular weight is twice the atomic weight, or, in other words, that the molecules consist of two atoms, Na-Na. Passing, next, to the products, we find only two, charcoal, and a substance called sodic oxide. As regards the last, we have every reason to believe that its molecules consist of two atoms of sodium united to a single atom of oxygen, Na_2O. About the charcoal molecules, we have absolutely no knowledge whatever; and we will, therefore, as is usual in such cases, represent them as consisting of single atoms. Hence, writing the products after the factors, we have—

CO_2	Na-Na		C	Na_2O.
Carbonic Dioxide.	Sodium.		Carbon.	Sodic Oxide.

Remembering, now, that the number of atoms on the two sides must be the same, it is evident that the amount of oxygen in a molecule of CO_2 will yield $2Na_2O$; and, further, that, to form two molecules of Na_2O, two molecules of Na-Na are necessary. Hence our reaction must be written:

$$CO_2 + 2Na\text{-}Na = C + 2Na_2O.$$

By this we learn that, from one molecule of carbonic dioxide (CO_2) and two molecules of sodium (2Na-Na), there are formed two molecules of sodic oxide ($2Na_2O$)

and one atom of carbon (C). It is probable that the atoms of carbon group themselves into molecules; but, as we know nothing about their constitution, we cannot express it by our symbols.

Both of the products of this process are solids, and will be found, at the close of the experiment, in the small iron crucible in which the sodium was melted and introduced into the jar of carbonic-dioxide gas. The sodic oxide is a white solid, which is very soluble in water, or, rather, combines with water to form what is called caustic soda, which dissolves in the liquid; and caustic soda, as you well know, is a very important chemical agent. But the chief interest in this experiment centres about the other product. Charcoal is one of the forms of carbon; and the peculiar chemical relations of this element, which are illustrated by our experiment, are not only highly interesting in themselves, but have an important bearing on the subject of these lectures. I shall, therefore, digress for a moment from my immediate topic, in order to bring these facts to your notice.

Carbon, as you probably know, is one of the most remarkable of the chemical elements. In the first place, it is most protean in the outward aspects which it assumes. These brilliant crystals of diamond, the hardest of all bodies; this black graphite, as extreme in softness as is the diamond in hardness; these still more familiar lumps of coal, are all formed of the same elementary substance. In the second place, the various forms of fuel used on the earth also consist chiefly of this element, which is, therefore, the great source of our artificial light and heat, and the reservoir of that energy which, by the aid of the steam-engine, man uses with such effect.

All carbonaceous materials used as fuel, whether wood, coal, oil, or gas, if not themselves visibly organized, were derived from organized structures, chiefly plants; and all the light, all the heat, all the power, which they are capable of yielding, were stored away during the process of vegetable growth. The origin of all this energy is the sun, and it is brought to the earth by the sun's rays. Coal is the charred remains of a former vegetation, and the energy of our coal-beds was accumulated during long periods in the early ages of the geological history of the earth. Wonderful as the truth may appear, it is no less certain that the energy which drives our locomotives and forces our steamships through the waves came from the sun, than that the water, which turns the wheels of the Lowell factories, came from the springs of the New-Hampshire hills. How it comes, how there can be so much power in the gentle influences of the sunbeam, is one of the great mysteries of Nature. We believe that the effect is in some way connected with the molecular structure of matter; but our theories are, as yet, unable to cope with the subject. That the power comes from the sun, we know; and, moreover, we are able to put our finger on the exact spot where the mysterious action takes place, and where the energy is stored; and that spot, singular as it may appear, is the delicate leaf of a plant.

This same carbonic dioxide, on which we are here experimenting, is the food of the plant, and, indeed, the chief article of its diet. The plant absorbs the gas from the air, into which it is constantly being poured from our chimneys and lungs, and the sun's rays, acting upon the green parts of the leaf, decompose it. The oxygen it contains is restored to the atmosphere, while the carbon remains in the leaf to form the structure of the growing plant. This change may be represented thus:

$$\underset{\text{Carbonic Dioxide.}}{CO_2} = \underset{\text{Carbon.}}{C} + \underset{\text{Oxygen.}}{O{=}O.}$$

Now, to tear apart the oxygen atoms from the carbon requires the expenditure of a great amount of energy, and that energy remains latent until the wood is burned; and then, when the carbon atoms again unite with oxygen, the energy reappears undiminished in the heat and light, which radiate from the glowing embers. Just as, when a clock is wound up, the energy which is expended in raising the weight reappears when the weight falls; so the energy, which is expended by the sun in pulling apart the oxygen and carbon atoms, reappears when those atoms again unite. This is one of the most wonderful and mysterious effects of Nature; for, although the process goes on so silently and unobtrusively as to escape notice, it accomplishes an amount of work compared with which most of the noisy and familiar demonstrations of power are mere child's-play. It is one of the greatest achievements of modern science, that it has been able to measure this energy in the terms of our common mechanical unit, the foot-pound; and we know that the energy exerted by the sun and rendered latent in each pound of carbon, which is laid away in the growing wood, would be adequate to raise a weight of five thousand tons one foot.

The chief interest connected with the experiment before us is to be found in the fact that it is almost the parallel to the process which is going on in the leaf of every plant that waves in the sunshine. Compare the two reactions as they are here written, the one over the other:

$$CO_2 + 2Na\text{-}Na = C + 2NaO.$$
$$CO_2 = C + O{=}O.$$

In the first, the cause of the breaking up of the CO_2 molecule is evident. The molecules of the sodium have what is called an intense affinity for the atoms of oxygen, and attract them with such power as to tear them away from the atom of carbon. Now, when you remember that the atoms of carbon and oxygen are united by such a force that it requires the great energy I have described to tear them apart, and in the light of this knowledge study the second reaction, you will fail to find in the symbols any adequate explanation of the effect. And they cannot explain it; for the sun's energy cannot be expressed by a chemical formula. But, yet, this energy does here precisely the same work which the sodium accomplishes in our crucible. Moreover, there is another striking analogy between the two processes, which must not be overlooked.

The carbonic dioxide is decomposed in a vegetable leaf; and, of the two products of the reaction, the oxygen gas escapes into the air, while the carbon is deposited in the vegetable tissue. This relation between the two products depends on the aëriform condition of oxygen on the one hand, and the great fixity of carbon on the other. Carbon is peculiar in this respect: In all its conditions, whether of diamond, graphite, or coal, it is one of the most fixed solids known. Even when exposed to the highest artificial heat, it never loses its solid condition, and so the molecules of carbon, as they form in the leaf, assume their native immobility, and become a part of the skeleton of the growing plant. To fully appreciate this remarkable relation of carbon to organic structures, you must recall the fact that the only other three elementary substances, of which animals and plants chiefly consist--oxygen, hydrogen, and nitrogen—are not only aëriform, but they are gases, which no amount of pressure or cold is able to reduce to the liquid or solid condition. All organized beings may be said to be skeletons of carbon, which have condensed around the carbon atoms the elements of water and of air.

This point is one of such interest that a familiar illustration of it may be acceptable. When a piece of wood is heated out of contact with the air, the volatile elements, hydrogen, oxygen, and nitrogen, are driven off in various combinations, while the carbon molecules are left behind, retaining the same relative position they had in the tree; and, if we examine the charcoal with a microscope, we shall find that it has preserved the forms and markings of the cells, and the rings of annual growth; and, in fact, all those details of structure which marked the kind of wood from which it was made.

My assistant has projected on the screen a magnified image of a thin section of wood, which has been thoroughly carbonized, and you see how strikingly the facts I have stated appear.

Now, just as the non-volatile carbon is deposited from the carbonic dioxide in the cell of the plant, so in our experiment is it deposited in the crucible. Both of the products of the reaction are to a great extent fixed, but the carbon by far the most so; and, in this experiment, all, or, at least, a great part, of the carbonic dioxide, which previously filled the jar, has deposited the carbon it contained in the iron crucible. In the plant the carbonic dioxide, which passes through the structure in the process of plant-life, leaves its carbon in the leaf or stalk; and so here, the carbonic dioxide, which is brought by the currents in the jar in contact with the heated sodium, leaves its carbon in the crucible. In order to show you that carbon has been thus formed, I will now remove the crucible, and quench it with water. The sodic oxide (Na_2O) dissolves, and the charcoal is set free, and you see that the water in this jar is black with the particles of floating charcoal.

Let us now pass on to study a remarkable series of chemical changes, in which carbonic dioxide also plays an important part. The first of the series is one with which you are all so familiar, that it is perhaps not important to repeat it in this place; but, as I am anxious that you should have the processes we are studying presented to you in visible form, I will make the trivial experiment of slaking some common lime.

The action is very violent, and great heat is developed. As we shall hereafter see, the evolution of heat is an indication of chemical combination, and, in the case before us, the lime unites with the water. Let us try to represent this change by our symbols.

Lime is a compound of a metal we call calcium and oxygen. It is, in a word, a metallic ore; and I have a small bit of the metal which it contains in this tube. By projecting an image of the tube on the screen, you can see almost all that I can, save only that the metal has a brilliant lustre and ruddy tint, like bismuth. A molecule of lime is formed of two atoms, one of this metal and the other of oxygen. Hence the symbol CaO. A molecule of water, as we know, is represented by H_2O. The product of the reaction is a light, white powder we familiarly call slaked lime, and its analysis, interpreted by its chemical relations, shows that it has the constitution CaO_2H_2. The chemical name is calcic hydrate, and the change by which it was produced we can now express thus:

$$\underset{\text{Lime.}}{CaO} + \underset{\text{Water.}}{H_2O} = \underset{\text{Calcic Hydrate.}}{CaO_2H_2}.$$

In this reaction, as you see, two molecules unite to form a third, which consists of the atoms of the other two. If, now, we mix this slaked lime with a larger body of water, the result is an emulsion called milk-of-lime, and consisting merely of particles of the hydrate suspended in water. A part of the hydrate actually dissolves; and, if we employ as much as 700 times its volume of water, the whole dissolves, forming a transparent solution. This milk-of-lime, then, is a solution of calcic hydrate, containing a large excess of the solid hydrate in suspension. But there is a very simple means of separating the solid from the solution.

We use for the purpose a circular disk of porous paper, called a filter, which we fold in the shape of a cone, and place in a glass funnel. On pouring the turbid liquid into the paper cone, the clear solution will trickle through the pores of the paper, but the solid sediment will be retained on the upper surface.

Having now obtained a clear solution of calcic hydrate ($CaO_2H_2 + Aq$), I propose to show you next the action of carbonic dioxide upon it.

For that purpose we will prepare some more of the gas, and, having poured our clear solution into this jar, we will pour in after it a quantity of carbonic dioxide, which, although a gas, is so heavy that we can handle it very much like a liquid. The gas is now resting on the solution, but the action is exceedingly slow; for, although the particles of the calcic hydrate are free to move in the liquid, and those of the carbonic dioxide in the space above the liquid, yet each is restricted to those spaces, and the two sets of molecules cannot come in contact, except at the surface of separation.

But, let us shake up the liquid, so as to bring the molecules of both liquid and gas in contact, and you see that, at once, we have a very marked change. The liquid becomes turbid, and, after a while, a quantity of a white powder will fall to the bottom, which, if collected and examined, will be found to be identical with chalk. Now that you are acquainted with our method of notation, I can best explain to you this change by writing at once the reaction:

$$\underset{\text{Calcic Hydrate.}}{(CaO_2H_2} + CO_2 + Aq.) = \underset{\text{Calcic Carbonate.}}{\underline{CaCO_3}} + (H_2O + Aq.).$$

The symbols of the factors of the reaction you will at once recognize, and you will also interpret the meaning of Aq., used to indicate that the calcic hydrate and carbonic dioxide come together in solution. Among the products of the reaction, the first symbol represents one molecule of calcic carbonate, the material of chalk. This body, being insoluble in water, drops out of the solution, and forms what is called a precipitate, a condition which we indicate arbitrarily by drawing a line under the symbol. The only other product of the reaction is water, which, of course, mingles with the great mass of water present, and this we express by $H_2O + Aq$.

I need not tell you that this white powder is not only the material of chalk, but the material of the limestone-rocks, which form so great a part of the rocky crust of our globe. Not only the rough mountain limestones, but the fine marbles, and that beautiful, transparent, crystalline mineral we call Iceland-spar, are aggregates of molecules, having the same constitution as those which have formed in this experiment. The differences of texture may, doubtless, be referred to differences of molecular aggregation; but we have not yet been able to discover, either what the difference is, or on what it depends.

In order to produce the last reaction, we poured the gas upon the solution of calcic hydrate; and the chalk was only produced as fast as the gas dissolved in the liquid. We shall obtain the reaction more promptly, if, instead of taking the gas itself, we employ a solution of the gas in water, previously prepared. Moreover, this form of the experiment will enable me to show you a phase of the process which might otherwise escape your notice. I need not tell you that we can easily obtain such a solution ready-made to our hands. That beverage, which we persist in miscalling soda-water, is simply an over-saturated solution of carbonic dioxide in water, made by forcing a large excess of the gas into a strong vessel filled with water. At the ordinary pressure of the air, water will dissolve its own volume of this gas; but, when forced in by pressure, the water dissolves an additional volume for every additional atmosphere of pressure. As soon, however, as this solution is drawn out into the air, the excess of gas above one volume escapes, causing the effervescence with which we are so familiar. Carbonic dioxide is formed in the process of fermentation by which beer and wine are prepared; and it is the escape of the excess of this gas, dissolved under pressure, which causes the effervescence of bottled beer and champagne. The solution in water (soda-water) is now supplied to the market in bottles called siphons, which are convenient for our purpose.

Notice that, as I permit the solution to flow into the lime-water, the same white powder appears as before; but, now, notice further that, as I continue to add the solution of carbonic dioxide, this white solid redissolves, and we have a beautifully clear solution. It is generally believed that, under these conditions, in presence of a great excess of carbonic dioxide, the molecule of calcic carbonate combines with additional atoms of carbon, oxygen, and hydrogen, to form the very complex molecule $H_2CaC_2O_6$, which is assumed to be soluble in water; but, as this point is one of doubt, I prefer to present the phenomenon to you as simply one of solution, and as illustrating a remarkable point in our chemical philosophy—the fact that the production of a given compound is frequently determined by the circumstance of its insolubility. The calcic carbonate forms, in the first instance, because this compound is insoluble; but, when a proper solvent like the aërated water is present in sufficient excess, no such compound results, or, at least, we have no evidence of its formation.

Most of my audience will be more interested, however, in this solution of chalk in soda-water (for such it is), from the fact that it plays a very important part in Nature, and is a common feature of domestic experience. Such a solution as this is what we call hard water, and spring-water is frequently in this condition. Such water is said to kill soap, and is disagreeable when used in washing, because the lime in solution forms with the fatty constituent of the soap an insoluble, sticky mass, which adheres to the hands or cloth. Moreover, when such water is boiled, the carbonic dioxide is driven off, and the water loses its power of holding the chalk in solution, which is deposited sometimes as a loose powder, but at other times as a hard crust on the sides of the boiler.

I cannot readily show you the reprecipitation under these conditions; but I have here a crust, which was formed in a steam-boiler in the manner I have described. A precisely similar action gives rise to the formation of stalactites in lime-caverns, and of a form of lime-rock called travertine. Some of the finest marbles have been formed in this way.

Thus it is that we have been imitating here the production of chalk, limestone, and marble, at least so far as the chemical process is concerned. The molecule of all these substances has the same constitution, expressed by the symbol $CaCO_3$. Now, it is evident that

$$\underset{\text{Calcic Carbonate.}}{CaCO_3} = \underset{\text{Lime.}}{CaO} + \underset{\text{Carbonic Dioxide.}}{CO_2}.$$

I mean simply by this, that it is theoretically possible to form, from one molecule of calcic carbonate, one molecule of lime and one molecule of carbonic dioxide; but it does not follow from this that it is practically possible to break up the molecule of calcic carbonate in this way; and we must avoid the error, not unfrequently made by chemical students, of being led astray by our notation. These equations, which we call reactions, are not like the equations of algebra. Any thing that can be deduced from an algebraic equation, according to the rules of the science, must be true; but it by no means follows that any combinations we may form with our symbols can be realized. We cannot deduce facts from chemical symbols. They are merely the language by which we express the *results of experiment;* and for this reason I have been, and shall be, very careful to show you the facts before I attempt to express them in chemical language. But, in the case before us, our caution is needless, for we can break up the molecule in the precise way which our assumed reaction indicates; and I will show you, lastly, two additional chemical processes, which will bring back our material to the condition of lime and carbonic dioxide, the substances from which we started.

The first is a reaction, identical with the one I have just written. Since the beginning of the lecture, I have been strongly heating some lumps of chalk in this platinum crucible. The process is a slow one; and it was necessary to begin the experiment early, in order that I might show you the result. The chemical change is identical, however, with that which may be observed in any lime-kiln, where lime is made by burning limestone. Each molecule of chalk, $CaCO_3$, looses a molecule of carbonic dioxide, CO_2, and we have left a molecule of lime, CaO. But the change in the appearance of the white mass produced by burning is so slight that I must bring in the aid of experiment to prove that any change has taken place; and, first of all, I must show you the test I am going to use.

In the first of these two jars I have an emulsion of chalk, and in the second milk-of-lime. Notice that this piece of paper, colored by a vegetable dye called turmeric, remains unchanged when dipped in the emulsion of chalk, but turns red in the milk-of-lime.

Let us test, now, the contents of our crucible. We will first empty it into some water. The white lumps almost instantly become slaked, and render the water milky. We will now dip in a sheet of turmeric-paper, and you see that, although we began with inactive chalk, we have obtained a material which acts on the turmeric-paper like caustic lime. Thus, then, we have regenerated the lime.

Let us next see if we can regenerate the carbonic dioxide:

In the last experiment, carbonic dioxide was produced, but it escaped so slowly, and in such small quantities, as entirely to escape notice. Where, however, limestone is burned on a large scale, the current of gas from the kiln is frequently very perceptible; and more than one poor vagrant, who has sought a night's lodging under the shelter of the stack, has been suffocated by the stream. But we can make evident the production of carbonic dioxide from chalk without the aid of such a sad illustration.

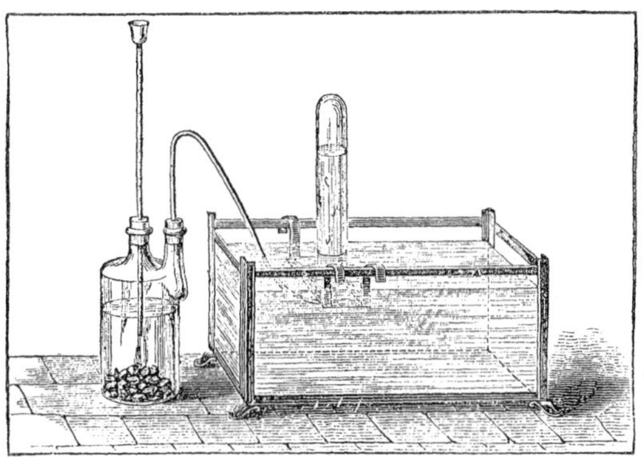

FIG. 22.—Pneumatic Trough, with Two-necked Gas-bottle.

In this bottle we have some bits of chalk. One of the two necks of the bottle is closed by a cork, through which passes tightly an exit-tube, to conduct away any gas that may be formed. The other is also corked, and through the cork passes a funnel-tube, by which I can introduce any liquid reagent into the bottle (Fig. 22). On pouring in some muriatic acid, a violent effervescence ensues, and a gas is formed which, flowing from the exit-tube, displaces the water in this glass bell.

The bell stands in what we call a pneumatic trough, and this simple apparatus for collecting gases must, I think, be familiar to all of my audience. The open mouth of the bell rests on the shelf of the trough under water, and the liquid is sustained in it by the pressure of the air. Let me, while the experiment is going on, write out the reaction:

$$\underset{\text{Chalk.}}{CaCO_3} + \underset{\text{Hydrochloric Acid.}}{(2HCl + Aq.)} = \underset{\text{Calcic Chloride.}}{(CaCl_2 + Aq.)} + \overline{CO_2}.$$

We already know the symbols of all the factors, and we may, therefore, confine our attention to the products.

The products are, first, carbonic-dioxide gas; and, secondly, a solution in water of a compound whose molecule consists of calcium and chlorine, and which we call calcic chloride. And, now that the jar is filled, I can easily show that we have regenerated carbonic dioxide. Removing the jar from the trough, we will first lower into it this lighted candle, and then pour into it some lime-water. The candle is instantly extinguished, and the lime-water rendered turbid.

Thus we end the torture of these molecules. You have seen how easily we have formed them, and how readily we have broken them up. We began with lime and carbonic dioxide, which we united to form chalk. We dissolved the chalk in a solution of CO_2, and learned how, in Nature, various forms of limestone could be crystallized from this solution. Lastly, we have recovered from the chalk the lime and carbonic dioxide with which we begun. I hope you have been able to follow these changes, and to understand the language in which they are expressed. If so, we have taken another step in advance, and, at the next lecture, shall be able to go on and classify these reactions, and thus prepare the way by which we may reach still further truth in regard to this wonderful microcosm of molecules and atoms.

Before, however, closing my lecture, I will embrace the opportunity offered by this division of my subject to explain, as briefly as I can, the principles of our chemical nomenclature. This nomenclature originated in 1787 with a committee of the French Academy of Sciences, a committee of which the great chemist Lavoisier was the ruling spirit. It was an attempt to indicate the composition of a substance by its name, and, for half a century after its adoption, it served most admirably the purpose for which it was devised, and exerted a marked influence on the development of chemistry. The nomenclature was based, however, on the dualistic theory, of which Lavoisier was the father, and, when at last our science outgrew this theory, the old names lost much of their significance and appropriateness. Within the last few years attempts have been made to modify the old nomenclature, so as to better adapt the names to our modern ideas. Unfortunately, the result, like most attempts to piece out an old garment, is far from satisfactory, and reviewers revel in the absurdities to which the nomenclature leads when applied to many of the products of modern chemical investigation. Fortunately, however, chemical symbols now supply to a great extent the place of philosophical names, and hence the nomenclature is a far less important feature in the new chemistry than it was in the old. I shall not, therefore, enter into much detail in regard to it, but limit myself to the statement of a few rules which will give you the key to the significance of the more common chemical terms.

The names of elementary substances are necessarily arbitrary. Those which were known before 1787 retain their old names, such as *sulphur*, *phosphorus*, *iron*, *gold*, and several others, including all the useful metals. Most of the more recently-discovered elements have been named in allusion to some prominent property, or some circumstance connected with their history: as *oxygen*, from ὀξὺς γεννάω (acid generator); *hydrogen*, from ὕδορ γεννάω (water generator); *chlorine*, from χλωρός (green); *iodine*, from ἰωδής (violet); *bromine*, from βρῶμος (fetid odor). The names of the newly-discovered metals have a common termination, *um*, as *potassium*, *sodium*, *platinum;* and, the names of several of the non-metallic elements end in *ine*, as *chlorine*, *bromine*, *iodine*, *fluorine*.

Passing next to binary compounds—that is, compounds of only two elements—we notice, first, that the simple compounds of the other elements with *oxygen* are all called *oxides*, and that, in order to distinguish the different oxides, we use adjectives formed from the name of the element with which the oxygen is combined, preferring however, in many cases, the Latin name to the English, both for the sake of euphony and in order to secure more general agreement in different languages. Thus we have—

Argentic oxide	Ag_2O
Plumbic oxide	PbO
Stannic oxide	SnO_2

When the same element forms with oxygen two compounds the termination *ic* is retained for the higher oxide, while the termination *ous* is given to the lower. Thus—

Ferr*ous* oxide	FeO
Ferr*ic* oxide	Fe_2O_3
Sulphur*ous* oxide	SO_2
Sulphur*ic* oxide	SO_3

If there are more than two oxides, or if, in any case, there are objections to the use of the termination *ous*, the necessary distinctions are made by means of Greek numeral prefixes:

Nitr*ous* oxide	N_2O
Nitr*ic* oxide	NO
Dinitr*ic* trioxide	N_2O_3
Nitr*ic* dioxide	NO_2
Dinitr*ic* pentoxide	N_2O_5
Carbon*ic* oxide	CO
Carbon*ic* dioxide	CO_2

The names of the *binary compounds* of the other elements are formed like those of the oxides.

Compounds of	Chlorine	are called	Chlor*ides*.
"	Bromine	"	Brom*ides*.
"	Iodine	"	Iod*ides*.
"	Fluorine	"	Fluor*ides*.
"	Sulphur	"	Sulph*ides*.
"	Nitrogen	"	Nitr*ides*.
"	Phosphorus	"	Phosph*ides*.
"	Arsenic	"	Arsen*ides*.
"	Antimony	"	Antimon*ides*.
"	Carbon	"	Carbon*ides*.

Moreover, the specific names in the several classes of compounds also follow the analogy of the oxides, thus:

Stann*ous* chloride	$SnCl_2$
Stann*ic* chloride	$SnCl_4$
Diferrous sulphide	Fe_2S
Ferrous sulphide	FeS
Ferric sulphide	Fe_2S_3
Ferric disulphide	FeS_2

And here, before we pass on to the names of compounds of a higher order, let me ask you to carefully fix in your memory the fact that the termination *ide* always indicates a compound containing only two elements.

Of compounds of three or more elements the most prominent class is that of the acids, bodies originally so called on account of their sharp or acrid taste. Now, the greater part of the inorganic or mineral acids are composed of the two elements hydrogen and oxygen, united to some third element, which is the characteristic constituent in each case; and, from this third element the acid takes its name, the terminations *ic* and *ous* being used as in the case of binaries to indicate a greater or less amount of oxygen in the compound. Thus we have—

Nitr*ous* acid	HNO_2
Nitr*ic* acid	HNO_3
Sulphur*ous* acid	H_2SO_3
Sulphur*ic* acid	H_2SO_4
Phosphor*ous* acid	H_3PO_3
Phosphor*ic* acid	H_3PO_4

In every acid we can by various chemical processes replace the hydrogen it contains with different metallic elements, and we thus obtain a very large class of compounds called salts. The generic name of the salts of each acid is formed by changing the termination *ic*, of the name of the acid, into *ate*, or the termination *ous* into *ite*, thus:

Sulphur*ous* acid	forms	Sulph*ites*,
Sulphur*ic* acid	"	Sulph*ates*,
Phosphor*ous* acid	"	Phosph*ites*,
Phosphor*ic* acid	"	Phosph*ates*,
Carbon*ic* acid	"	Carbon*ates*,
Silic*ic* acid	"	Silic*ates*,

and the different salts of the same acid are distinguished by adjectives as before. For example:

Nitric acid	HNO_3
Sodic nitrate	$NaNO_3$
Potassic nitrate	KNO_3
Argentic nitrate	$AgNO_3$

So also:

Sulphuric acid	H_2SO_4
Potassic sulphate	K_2SO_4

Calcic sulphate........................	$CaSO_4$
Mercur*ous* sulphate......................	Hg_2SO_4
Mercur*ic* sulphate......................	$HgSO_4$
Ferr*ous* sulphate......................	$FeSO_4$
Ferr*ic* sulphate........................	$Fe_2(SO_4)_3$

The terminations *ous* and *ic*, used in the names of these salts, indicate the same difference in the condition of the metallic element which determines the union of the metal with more or less oxygen. Ferrous and ferric sulphates, for example, correspond to ferrous and ferric oxides. The nature of this difference will be discussed in the chapter on quantivalence.

There is an important class of compounds which bears to water a relation similar to that which salts sustain to their respective acids. This class of compounds is called the *hydrates*, and may be regarded as derived from water, by replacing one-half of its hydrogen. Thus we have—

Potassic hydrate............	KOH	from	HOH
Calcic hydrate..............	CaO_2H_2	"	$2HOH$
Bismuthic hydrate..........	BiO_3H_3	"	$3HOH$
Silicic hydrate..............	SiO_4H_4	"	$4HOH$

So also:

Ferr*ous* hydrate........................	FeO_2H_2
Ferr*ic* hydrate........................	$Fe_2O_6H_6$

The very interesting theoretical relations of the hydrates will hereafter be discussed.

When the hydrogen of an acid is only in part replaced, or is replaced by more than one metallic element, the constitution of the resulting salt may still be indicated by the name, as in the following examples:

Hydro-disodic phosphate.............	H,Na_2PO_4
Potassio-aluminic sulphate...........	$K_2Al_2(SO_4)_4$

In like manner the relative proportions of the several ingredients of a salt may be indicated, as in—

Tetrahydro-calcic diphosphate....... $H_4Ca(PO_4)_2$
Disodic tetraborate (borax).......... $Na_2B_4O_7$

But, as is evident, names like the last two are practically useless, and, when we attempt to extend the nomenclature to organic compounds, we are led into still greater absurdities; so that, although by giving arbitrary names to various groups of atoms called compound radicals we have been able, to a limited extent, to adapt the nomenclature to this class of substances, yet we have been compelled in many cases to resort to trivial names like those used before the adoption of the nomenclature. The names oil of vitriol, corrosive sublimate, calomel, saltpetre, borax, cream-of-tartar, etc., of the last century, have their counterparts in aldehyde, glycol, phenol, urea, morphine, naphthaline, and many other familiar names of our modern science. Of course, such names are subject to no rules, and, although they have been usually selected with care, and indicate by their etymology important relations or qualities, they must be associated separately with the substances they designate.

LECTURE VIII.

CHEMICAL CHANGES CLASSIFIED.

Among chemical reactions we may distinguish three classes: 1. Those in which the molecules are broken up into atoms; 2. Those in which atoms are united to form molecules; and, 3. Those in which the atoms of one molecule change places with those of another. Reactions of the first kind are called *analysis*, those of the second *synthesis*, and those of the third *metathesis* —terms derived from the Greek, and signifying respectively to *tear apart*, to *bind together*, and to *interchange.*

This classification is one of great theoretical importance. But it must be further stated that a simple analytical or synthetical reaction, as here defined, is seldom if ever realized in Nature. Almost every chemical process is attended both with the breaking up of molecules into atoms and the regrouping of these atoms to form new molecules, that is, it involves both analysis and synthesis; and this is true even in the many cases where the products or factors of the chemical reaction are elementary substances; for, when the molecules of the elementary substances consist of two or more atoms, the breaking apart or coalescing of these atoms, although they are atoms of the same element, constitutes analysis or synthesis, as here defined. Thus, when, in the burning of hydrogen gas, this elementary substance unites with the oxygen of the air to form water, the molecules of oxygen must be divided into atoms before the synthesis of the water molecule is possible; and so, on the other hand, when water is decomposed, the resulting atoms of oxygen unite by twos to form molecules of oxygen gas; and this pairing is, according to our definition, a process of synthesis. The chemical reactions, which express these changes, illustrate very clearly the point here made:

Burning of Hydrogen Gas.

$$2H\text{-}H + O\text{=}O = 2H_2O.$$

Hydrogen Gas. Oxygen Gas. Water.

Decomposition of Water.

$$2H_2O = 2H\text{-}H + O\text{=}O.$$

Water. Oxygen Gas. Hydrogen Gas.

The first states that from one molecule of oxygen are formed two molecules of water, and this, of course, necessitates a division of the oxygen molecules; while the second states that from two molecules of water only one molecule of oxygen gas results, a process which involves the union of the two oxygen atoms, previously separated in the two molecules of water. Indeed, a purely analytical or a purely synthetical reaction would only be possible theoretically in those cases where elementary substances were involved, whose molecules consist of a single atom, that is, where the molecule and the atom are identical, and we can recall no well-defined reactions of this kind.

But, although we should be obliged to seek among the unfamiliar facts of chemistry for examples of pure analysis or pure synthesis, yet processes, in which one or the other is the predominant feature, and which illustrate the special characteristics of each, are close at hand. Some of these I now propose to bring before you, beginning with the analytical processes, and I shall select such examples as incidentally illustrate important principles, or interesting facts, of the science.

Afterward, we will pass to the metathetical reactions, which are not only very common, but constantly occur undisturbed by other modes of chemical change; and the study of this very important class of phenomena will show us some of the latest phases which our chemical philosophy has assumed.

Of the analytical reactions I will select for our first illustration the process by which oxygen gas is usually made. The common source of oxygen is a white salt, now well known under the name of chlorate of potash, but which, in the nomenclature of our modern chemistry, is called potassic chlorate. I presume, if we should inquire into the cause of the present notoriety of this chemical preparation, we should find that it owed its reputation to the chlorate of potassa troches, and there is no doubt that, when judiciously used, this salt has a very soothing effect on an irritated throat. But, after all, the great mass of the potassic chlorate manufactured is used for fireworks or for making oxygen gas, and it is to the last use we now propose to apply it. For this purpose, we have only to heat the salt to a low, red heat in an appropriate vessel. We use here a copper flask, and connect the exit-tube with the now familiar pneumatic trough. While my assistant is preparing the oxygen gas, I will explain to you the process.

Although potassic chlorate is a non-volatile solid, and we have no direct means of weighing its molecules, yet, from the purely chemical evidence we possess, there is no doubt whatever about its molecular constitution. It is expressed by the symbol $KClO_3$, and, in the process before us, the potassic chlorate simply breaks up into another salt called potassic chloride and oxygen gas,

$$\underset{\text{Potassic Chlorate.}}{KClO_3} = \underset{\text{Potassic Chloride.}}{KCl} + \underset{\text{Oxygen Atoms.}}{O_3},$$

that is, each molecule of the salt gives a molecule of potassic chloride and three atoms of oxygen. Notice that I say three atoms; for this is a point to which I must call your attention.

We are not dealing here with an example of pure analysis, although that feature of the reaction predominates over every other. Oxygen *gas* is the product formed; and, as I have several times said, we know that the molecules of oxygen consist of two atoms. Hence, the three atoms which the heat drives off must pair, and, from three atoms, we can only make one molecule. What, then, is to become of the third atom, which seems to be left out in the cold? You must have already answered this question; for you remember that our symbols only express the change in one of the many millions of molecules which are breaking up at the same instant; so there can be no want of a mate for our solitary atom. Indeed, two molecules of chlorate will give us just the number of atoms we want to make three molecules of oxygen gas. Hence, we should express the change more accurately by doubling the symbols:

$$\underset{\text{Potassic Chlorate.}}{2KClO_3} = \underset{\text{Potassic Chloride.}}{2KCl} + \underset{\text{Oxygen Gas.}}{3O{=}O}.$$

Let me next remind you that these symbols express exact quantitative relations; and, as some of my young friends may desire to know how to calculate the amount of chlorate they ought to use in order to make a given volume, say, ten litres of oxygen, I will, even at the risk of a little recapitulation, go through the calculation: A molecule of $KClO_3$ weighs $39.1+35.5+48 = 122.6$ m.c., and two molecules will weigh 245.2 m.c. These yield 2KCl, weighing $2(39.1+35.5) = 149.2$ m.c., and 3O=O, weighing 96 m.c. We must next find the weight of ten litres of oxygen gas. To find the weight of one litre we multiply the specific gravity of the gas, or half molecular weight, by $\frac{9}{100}$. Now, $\frac{9}{100} \times 16 = 1.44$ gramme. Hence, ten litres weigh 14.4 grammes. But, if 96 m.c. of gas are made from 245.2 m.c. of salt, then 14.4 grammes would be obtained from a quantity easily found from the proportion:

$$96 : 245.2 = 14.4 : x = 36.78 \text{ grammes.}$$

I think, after this, we will assume that these quantitative relations are all right, and let them take care of themselves. Returning to the experiment, before I show that the products are those which I have described, let me give just a word of caution to any of my young friends present, who may like to repeat it.

We find that it is best to mix our chlorate with a heavy black powder, known in commerce as black oxide of manganese. What the effect of the powder is we do not know, for it is wholly unchanged in the process. But, in some way or other, it eases off the decomposition, which is otherwise apt to be violent. In buying the black oxide of manganese you must take care that it has not been adulterated with coal-dust—for a mixture of coal-dust and chlorate explodes with dangerous violence when heated, and serious accidents have resulted from the cupidity which led to such adulteration. Let me, moreover, say in general that, although I highly approve of chemical experiments, as a recreation for boys, they ought always to be made under proper oversight, and according to exact directions, and I would warmly recommend, as a trustworthy companion for all beginners, the abridgment of "Eliot and Storer's Manual of Chemistry," recently edited by Prof. Nichols, of the Institute of Technology.

But how shall I show you that this gas we have obtained is oxygen? I know of no better way than to test it with one of our watch-spring matches. . . . In no other gas will iron burn like this.

So much for the oxygen. Let us next turn to the other product, that I called potassic chloride. This is left in the retort, forming a solid residue, but, as it would take a long time to bring what we have just made into a presentable condition, we must be content to see some of the product of a former process, which I have in this bottle.

At a distance, you cannot distinguish the white salt from the potassic chlorate with which we started, but, if you compared the two carefully, you would see that there was a very great difference between them. I can only show you that the crystals of the two salts have wholly different forms. For this purpose I have crystallized them on separate glass plates, and I will now project a magnified image of the crystals on the screen. There you see them beautifully exhibited on the two illuminated disks side by side. The square figures on the left-hand disk (Fig. 23) are the projections of the cubes of *potassic chloride*, which differ utterly in form from the rhombic plates of *potassic chlorate* that appear on the right (Fig. 24).

The second example of an analytical process which I have to show you is also familiar to many of my audience, and cannot fail to be interesting to the rest; for it is the process by which nitrous oxide is prepared, the gas now so much used by the dentists as an anæsthetic. It was formerly called laughing-gas, but the peculiar intoxication it causes, when inhaled under certain conditions, has been almost forgotten in its present

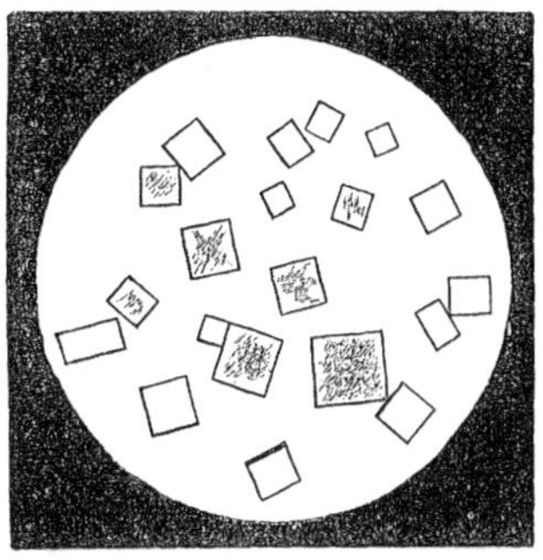

FIG. 23.—Crystals of Potassic Chloride.

FIG. 24.—Crystals of Potassic Chlorate.

beneficent application in minor surgery. Nitrous oxide is made from a well-known white salt, prepared from one of the secondary products of the gas-works, and called nitrate of ammonia, or ammonic nitrate. When this salt is gently heated in a glass flask, its molecules split up into those of nitrous oxide and water.

Again, let us make use of the time required for the experiment to explain the process. The molecules of ammonic nitrate have the constitution $N_2H_4O_3$, and the change may be represented thus:

$$\underset{\text{Ammonic Nitrate.}}{N_2H_4O_3} = \underset{\text{Water.}}{2H_2O} + \underset{\text{Nitrous Oxide.}}{N_2O.}$$

The experiment has been arranged so as to show both of the products (Fig. 25). The water condenses in this test-tube, while the gas passes forward, and is collected over a pneumatic trough. But what evidence can I give you that these are, in fact, the products? As regards the water, you would readily recognize the familiar liquid, which has collected in the tube, could you examine and taste it. But, as I cannot offer you this evidence, I will seek for another. Most of you must be familiar with the remarkable action of the alkaline metals on water. You see how this lump of potassium inflames the moment it touches the liquid.

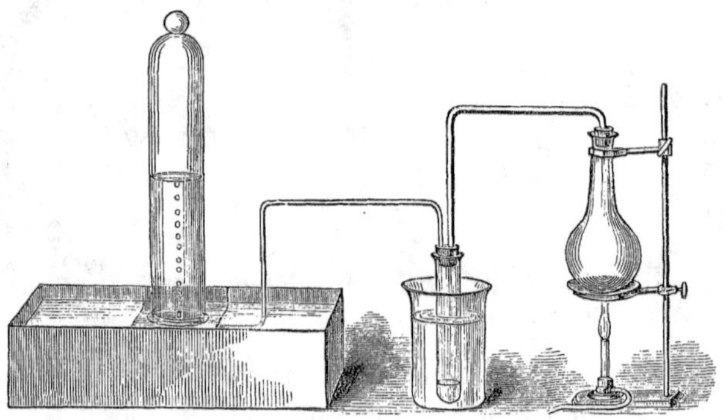

FIG. 25.—Preparation of Nitrous Oxide and Water, from Ammonic Nitrate.

Let us now see whether it will act in a similar way on the liquid which has condensed in our tube. . . . There can be no doubt that we are dealing with water. Next for the gas. Nitrous oxide has the remarkable quality, not only of producing anæsthesia, but also of sustaining the combustion of ordinary combustibles with great brilliancy—like oxygen gas. But there is a marked difference between nitrous oxide and oxygen, which an experiment will serve to illustrate, and this, at the same time, will show us that the gas we have obtained in our experiment is really nitrous oxide.

Taking a lump of sulphur, I will, in the first place, ignite it, and when it is only burning at a few points I will immerse it in a jar of oxygen. As you see, it at once burns up with great brilliancy. Taking now a similar lump of sulphur, and waiting until you all admit that it is ignited more fully than before, I will plunge it into this jar of gas we have just prepared, and which we assume to be nitrous oxide. . . . It at once goes out, and the reason is obvious. There is an abundance of oxygen in the nitrous oxide—relatively, more than twice as much as in the air; but, in the molecules of N_2O, the oxygen atoms are bound to the atoms of nitrogen by a certain force, which the sulphur at this temperature is unable to overcome. Let me, however, heat the sulphur to a still higher temperature, until the whole surface is burning, and you see that it burns as brilliantly in the compound as it does in the elementary gas.

The last example of an analytical reaction, which we shall have time to examine, is furnished by a remarkable compound of iodine and nitrogen, called iodide of nitrogen. Iodine is an elementary substance, resembling chlorine, which is extracted from kelp, that common broad-leafed sea-weed abounding on our coast. It is a very volatile solid, and gives a violet-colored vapor, whence its name from the Greek word ιώδης. When heated gently with aqua ammonia, the iodine takes from the ammonia a portion of nitrogen, and forms with it a very explosive compound whose molecule has the constitution NI_3. We have prepared a small quantity of the substance, and the black powder is now resting on this anvil, wrapped in filtering-paper. The slightest friction is sufficient to determine the breaking up of these very unstable molecules, and the decomposition of the compound into iodine and nitrogen. A mere touch with a hammer is followed by a loud report, when you notice a cloud of violet vapor, which indicates that the iodine has been set free:

$$\underset{\text{Iodide of Nitrogen.}}{2NI_3} = \underset{\text{Nitrogen Gas.}}{N\equiv N} + \underset{\text{Iodine-Vapor.}}{3I\text{-}I.}$$

In this case, as in previous examples, the atoms, when liberated, unite in pairs to form molecules of nitrogen gas on the one side, and molecules of iodine-vapor on the other; and, since a single molecule does not yield an even number of atoms of either kind, we double the symbols.

There is one characteristic of analytical reactions which must be carefully noticed. The parting of atoms (and it must be remembered that by an analytical reaction we merely mean this phase of a chemical process) is attended by the absorption of heat; although—as in the last experiment—the effect is often masked by other causes. But this truth can only be made evident by comparing the results of careful measurements, which cannot be made rapidly, and whose discussion, even, would be out of place at this time. I must, therefore, content myself with stating the fact as one of the definite results of science, and pass on to some examples of synthesis—reactions of the opposite class.

One of the most striking illustrations of the direct union of two molecules, to form a third, is furnished by the action of ammonia gas on hydrochloric-acid gas. Without entering into any details in regard to the processes by which these two aëriform substances are prepared, let it be sufficient to say that, in the glass flask on the right-hand side of this apparatus (Fig. 26), are the materials for making hydrochloric acid, and in the similar flask on the left those for making ammonia. The exit-tubes from these flasks deliver the two gases into this large glass bell, where they meet, and the chemical reaction takes place. The reaction is very simple, and one in regard to which we have no doubt, for the molecules of both of the factors have been weighed and analyzed. It is exprèssed thus :

NH_3	+	HCl	=	NH_4Cl.
Ammonia Gas.		Hydrochloric-Acid Gas.		Ammonic Chloride.

FIG. 26.—Combination of Ammonia and Hydrochloric-Acid Gases.

As you see, the atoms of a molecule of ammonia unite with those of a molecule of hydrochloric acid to form a single molecule of ammonic chloride, and, although the reaction may imply the breaking up, to a certain extent, of the molecules of the two factors, yet the subsequent synthesis is the chief feature. Ammonic chloride is a solid, and the sudden production, from two invisible gases, of the white particles of this salt, which fill the bell with a dense cloud, is a very striking phenomenon.

The second example of synthesis I have chosen is equally striking. Here, also, the factors of the reaction are both gases.

The lower jar (Fig. 27) contains a gas called nitric oxide, like nitrous oxide, a compound of oxygen and nitrogen, but containing a relatively larger proportion of oxygen. Its molecule has the constitution NO.

The upper jar contains oxygen, and, on removing the thin glass which now separates the two gases, you notice an instantaneous change. A deep-red vapor soon fills the glass. This red product is still another compound of nitrogen and oxygen, called nitric peroxide, whose symbol is NO_2, and the reaction is simply this:

$$\underset{\text{Nitric Oxide.}}{2NO} + O{=}O = \underset{\text{Nitric Peroxide.}}{2NO_2.}$$

Here a molecule of nitric oxide takes only an atom of oxygen, and, since each molecule of oxygen gas consists of two atoms, it will supply the need of two molecules of NO.

FIG. 27.—Combination of Nitric-oxide and Oxygen Gas.

Since the two factors and the single product of this process are all gases, the reaction before us is well adapted to illustrate another fact in regard to our symbols, of which I have not as yet directly spoken. If, in writing reactions, care is taken that each term shall always represent one or more perfect molecules, so far as their constitution is known—then the symbols will always indicate, not only the relative weights, but also the relative volumes of the several factors and products when in the state of gas. That this must be the case, you will see when you remember that equal volumes of all gases under the same conditions have the same number of molecules, and hence that all gas-molecules have the same volume. The symbol of one molecule represents what we will call a unit volume, and the number of these unit volumes concerned in any reaction is the same as the number of molecules. We can read the reaction before us thus: Two volumes of nitric-oxide and one volume of oxygen gas yield two volumes of nitric peroxide.

Three volumes, therefore, become two. If this is the case, there must be a partial vacuum in the jar, and, on opening the stop-cock, you hear the whistle which the current of air produces as it rushes in to establish an equilibrium.

We come now to still another example of a synthetical reaction, and, to illustrate this, the apparatus before you has been prepared (Fig. 28). The metallic leaf in the upper of the two glass jars is made of brass, which consists of the two metals, zinc and copper. In the lower jar we have chlorine gas. The air has been exhausted from the upper jar by a pump, and, on opening the stop-cock, the chlorine gas will rush in from the lower jar to take its place. Chemical union at once results, and notice the appearance of flame, which is an indication that great heat is produced by this chemical change. The change here is very simple. The atoms of chlorine unite directly with the atoms both of zinc and of copper, forming two compounds, which we call respectively zincic chloride, and cupric chloride. One reaction will serve for both metals, as the two are similar, differing only in the symbols of the metals. Take copper—

FIG. 28.—Union of Chlorine with Tinsel.

$$\underset{\text{Copper.}}{Cu} + \underset{\text{Chlorine Gas.}}{Cl\text{-}Cl} = \underset{\text{Cupric Chloride.}}{CuCl_2}.$$

As in analytical reactions heat is absorbed, so in synthetical reactions heat is evolved. You were all witnesses of the fact that heat was evolved in this last reaction, and it is equally true that heat was developed in each of the two previous experiments. In the combination of ammonia with hydrochloric acid (Fig.

26), this fact was made evident by the thermometer we placed in the bell for the purpose, and in the combination of nitric oxide with oxygen by the initial expansion which attended the first union of the two gases, and which would have lifted off the upper bell (Fig. 27), had I not firmly held it in its place. As, however, the product rapidly cooled to the temperature of the air, the initial expansion was soon followed by the condensation to which I called your attention.

Now, the principle illustrated by these three experiments is universally true, and the point is so important that I will make still another experiment in order to illustrate this feature of synthetical reactions still further. In this glass I have placed a small piece of phosphorus, and now I will drop upon it a few crystals of iodine. Direct combination between the phosphorus and iodine at once takes place, and the heat developed by this union is sufficient to inflame the uncombined phosphorus which I have intentionally added in excess.

The principle here illustrated is one of the greatest importance in the theory of chemistry, and this class of phenomena has been the object of extended investigation. Not only has it been shown that the principle here stated is in general true, but also that the amount of heat liberated by the union of the same atoms to form the same molecules is always constant, and this amount has, in very many cases, been measured. It has further been proved by actual experiment that, when, by any cause, the atoms thus joined are separated, exactly the same amount of heat is absorbed. In chemical processes, where, as a general rule, there are both analysis and synthesis, the thermal relations depend primarily on the extent to which these two effects neutralize each other; but changes in the state of aggregation, and other physical causes, constantly intervene to modify the result.

There is one class of chemical processes in which the thermal effects are so great, so striking, and so important, as to subordinate all other phenomena. I refer to the common processes of combustion, on which we depend for all our artificial light and heat. To these processes I shall next ask your attention, for, although they are only further illustrations of the principle just stated, yet, they play such an important part in Nature, and have been so often the battle-ground between rival chemical theories, that they demand our separate attention. I will open the subject by burning in the air a piece of phosphorus.

Before this intelligent audience it is surely unnecessary to dwell on the elementary facts connected with the class of phenomena of which this is the type. It will only be necessary for me to call to your recollection the main points, and then to pass to the few features which I desire especially to illustrate. In regard to the main points, no experiment could be more instructive than this. This large glass jar is filled with the same atmospheric air in which we live. Of this atmospheric air one-fifth of the whole material consists of molecules of oxygen gas in a perfectly free and uncombined condition; for, although they are mixed with molecules of nitrogen gas, in the proportion of four to one, and, although the presence of this great mass of inert material greatly mitigates the violence of our ordinary processes of burning, it does not, in any other respect, alter the chemical relations of the oxygen gas to combustible substances. These combustibles are, for the most part, compounds of a few elements—carbon, hydrogen, sulphur, and phosphorus—including the elementary substances themselves, and our common combustibles are almost exclusively compounds of hydrogen and carbon only. Their peculiar relations to the atmosphere depend solely on the fact that the atoms of these bodies attract oxygen atoms with exceeding energy, and it is only necessary to excite a little molecular activity in order to determine chemical union between the two. This union is a simple synthetical reaction, and, like all processes of that class, it is attended with the liberation of heat. The chief feature which distinguishes the processes of burning from other synthetical reactions is the circumstance that the heat generated during the combination is sufficient to produce ignition—in other words, to raise the temperature of the materials present to that point at which they become luminous, and the brilliant phenomena which thus result tend to divert the attention from the simple chemical change, of which they are merely the outward manifestation. In the case of our ordinary combustibles, the real nature of the process is still further obscured by the additional circumstance that the products of the burning—carbonic dioxide and aqueous vapor—are invisible gases, which, by mixing with the atmosphere, so completely escape rude observation that their existence even was not suspected until about a century ago, when carbonic dioxide was first discovered by Dr. Black. Although these aëriform products necessarily contain the whole material, both of the combustible and of the oxygen with which the combustible has combined, there is a seeming annihilation of the combustible, which completely deceived the earlier chemists. In the case before us, however, the product of the combustion is a solid, and it is this circumstance which makes the experiment so instructive. Almost every step of the process can be here seen. You noticed that we lighted the phosphorus in order to start the combustion—for this combustible, like every other, must be heated to a certain definite temperature before it bursts into flame. This temperature is usually called the point of ignition, and differs greatly for different combustibles. While phosphorus inflames below the temperature of boiling water, coal and similar combustibles require a full red heat. If, as our modern theory assumes, increased temperature merely means an increased velocity of molecular motion, the explanation of these facts would seem to be that a certain intensity of molecular activity is necessary in order to bring the molecules of oxygen sufficiently near to those of the combustible to enable the atoms to unite, and that the point of ignition is simply the temperature at which the requisite molecular momentum is attained. But the process once started continues of itself, for it is a characteristic of those substances we call combustible that, as soon as a part of the body is inflamed, the heat developed by the chemical union is sufficient to maintain the temperature of the adjacent mass at the ignition-point.

Passing next to the chemical process itself, nothing could be simpler than the change which is taking place in the experiment before us. It is an example of direct synthesis. This white powder which you see falling in such abundant flakes is the solid smoke of this fire. It is formed by the union of the phosphorus and oxygen—two atoms of phosphorus uniting with five of oxygen to form a molecule of this solid, which we call phosphoric oxide, and whose symbol we may write thus, P_2O_5.

But, neither the conditions of the burning nor the chemical change itself, although so beautifully illustrated here, are nearly so prominent facts as the manifestation of light and heat, which attends the process; and these brilliant phenomena wholly engrossed the attention of the world until comparatively recently, and indeed they still point out what is really the most important circumstance connected with this class of phenomena. The union of combustible bodies with oxygen is attended with the development of an immense amount of energy, which takes the form of light or heat, as the case may be. Moreover, it is also true that the amount of energy thus developed depends solely on the amount of combustible burnt, and not at all on the circumstance that the burning is rapid or slow. Thus, in the case before us, the amount of heat developed by the burning of an ounce of phosphorus is a perfectly definite quantity, and would not be increased if the combustion were made vastly more intense. So it is with other combustibles. The table before you gives the amount of energy developed by the burning of one pound of several of the more common combus-

Calorific Power from One Pound of Each Combustible.

	English Units of Heat.	Foot-pounds.
Hydrogen	62,032	47,888,400
Marsh-gas	23,513	18,152,350
Olefiant gas	21,344	16,477,880
Wood-charcoal	14,544	11,228,000
Alcohol	12,931	9,982,890
Sulphur	4,070	3,141,886

tibles, estimated, in the first place, in our common units of heat, and, in the second place, in foot-pounds. But, although the amount of energy is thus constant, depending solely on the amount of the combustible burnt, the brilliancy of the effect may differ immensely. A striking illustration of this fact I can readily show you.

For this purpose I will now repeat the last experiment, with only this difference, that, instead of burning the phosphorus in air, I will burn the same amount as before in a globe filled with pure oxygen.* We shall, of course, expect a more violent action, because, there being here no nitrogen-molecules, there are five times as many molecules of oxygen in the same space. Hence, there are five times as many molecules of oxygen in contact with the phosphorus at once, and five will combine with the phosphorus in the same time that one did before. But, with this exception, all the other conditions of the two experiments are identical. We have the same combustible, and the same amount of it burnt. We have, therefore, the same amount of energy developed, and yet how different the effect! Phosphorus burns brightly even in air, but here we have vastly greater brilliancy, and the intensity of the light is blinding.

What is the cause of the difference? One obvious explanation will occur to all: The energy in this last experiment has been concentrated. Although only the same amount of heat is produced in the two cases, yet, in the last, it is liberated in one fifth of the time, and the effect is proportionally more intense. The intensity of the effect is shown simply in two circumstances: first, a higher temperature; and, secondly, a more brilliant light. Of these, the first is fully accounted for in the explanation just suggested; for, if five times as much heat is liberated in a given time, it must necessarily raise the temperature of surrounding bodies to a much higher degree. I need not go beyond your familiar experience to establish this principle, although temperature is a complex effect, depending, not only on the amount of heat liberated, but also on the nature of the material to be heated, and on conditions which determine the rapidity with which the heat is dissipated. But the matter of the light is not so obvious. Why should more rapid burning be attended with more brilliant light? It is so in the present case; but is it always so? We can best answer this question by a few experiments, which will teach us what are the conditions under which energy takes the form of light; but these experiments we must reserve until the next lecture.

LECTURE IX.

THE THEORY OF COMBUSTION.

As our last hour closed, we were studying the phenomena of combustion. I had already illustrated the fact that, so far as the chemical change was concerned, these processes were examples of simple synthesis, consisting in the union of the combustible atoms with the oxygen atoms of the air, and that the sole circumstance which distinguished these processes from other synthetical reactions was the amount of energy developed. There were three points to which I directed your attention in connection with this subject: 1. The condition of molecular activity, measured by the temperature or point of ignition, which the process requires. 2. The chemical change itself, always very simple. 3. The amount of energy developed, and the form of its manifestation. This last point is the phase of these phenomena which absorbs the attention of beholders, and the one which we have chiefly to study. I stated in the last lecture that the amount of energy developed depended solely on the nature and amount of the combustible burnt, but I also showed that both the intensity and the mode of manifestation of this energy varied very greatly with the circumstances of the experiment. The intensity of the action we traced at once to the rapidity of the combustion, but the conditions which determine whether the energy developed shall take the form of heat or light we have still to investigate, and no combustible is so well adapted as hydrogen gas to teach us what we seek to know.

Here, then, we have a burning jet of hydrogen. It is not best for me to describe, in this connection, either the process or the apparatus by which this elementary substance is made, and a constant supply maintained at the burner, as I wish now to ask your attention exclusively to the phenomena attending the burning of the gas; and let me point out to you, in the first place, that hydrogen burns with a very well-marked flame. The flame is so slightly luminous that I am afraid it cannot be seen at the end of the hall, but I can make it visible by puffing into it a little charcoal-powder.

Now, all gases burn with a flame, and flame is simply a mass of gas burning on its exterior surface. As the gas issues from the orifice of the burner, the current pushes aside the air, and a mass of gas rises from the jet. If the gas is lighted—that is, raised to the point of ignition—this mass begins to combine with the oxygen atoms of the air at the surface of contact, and the size of the flame depends on the rapidity with which the gas is consumed as compared with the rapidity with which it is supplied. By regulating the supply with a cock, as every one knows, I can enlarge or diminish the size at will.

The conical form of a quiet flame results from the circumstance that the gas, as it rises, is consumed, and thus the burning mass, which may have a considerable diameter near the orifice of the jet, rapidly shrinks to a point as it burns in ascending.

But we must not spend too much time with these

details, lest we should lose sight of the chemical philosophy, which it is the main object of this course to illustrate. The chemical change here is even more simple than in the experiment with phosphorus, and consists solely in a direct union of the hydrogen atoms of the gas with the oxygen atoms of the air. Indeed, in another connection, we studied the reaction at an early stage in this course of lectures; when, in order to illustrate the characteristic feature of chemical combination, we exploded a mixture of hydrogen and oxygen gases. The reaction obtained under those conditions was identical with that here. We had not then learned to express the chemical change with symbols; but now I may venture to write the reaction on the black-board:

$$\underset{\text{Hydrogen Gas.}}{2H\text{-}H} + \underset{\text{Oxygen Gas.}}{O{=}O} = \underset{\text{Steam.}}{2H_2O.}$$

It would be very easy to show you that, as the symbols indicate, from two volumes of hydrogen, and one of oxygen, two volumes of steam are formed; but the experiment requires a great deal of time, and the result could not readily be made visible to this audience. I must content myself with proving that water is really produced by the hydrogen flame.

The apparatus we use looks complicated, but is, in fact, very simple (Fig. 29). By means of an aspirator the products of combustion are sucked through a long glass tube, which is kept cool by a current of water in a jacket outside. The flame burns under the open and flaring mouth of the tube, and the liquid, which condenses, drops into a bottle at the other end.

You must not expect that any considerable amount of water can be produced in this way. In the union of the two gases to liquid water, a condensation of 1,800 times takes place, so that, in order to obtain a quart of liquid water, we must burn 1,200 quarts of hydrogen gas, and take from the air 600 quarts of pure oxygen; and this, on the scale of our experiment, would be a very slow process. We have here obtained barely an ounce of liquid, although the jet has been burning for more than an hour. In order to show that the product is really water, I will apply the same test I used in a former experiment. We will pour the liquid into a shallow dish, and drop upon it a bit of potassium. . . . The hydrogen-flame, which at once bursts forth, gives the evidence we seek.

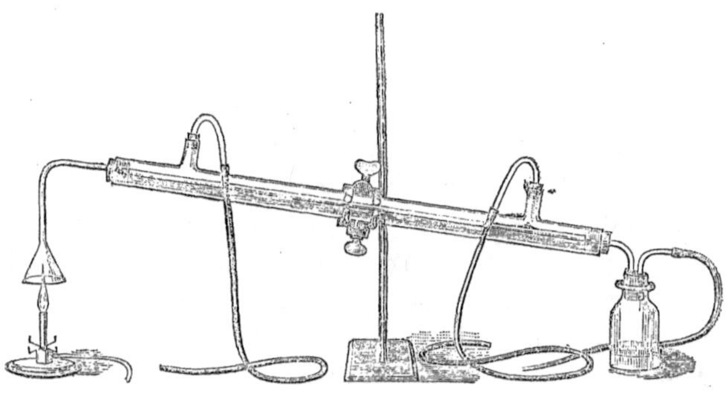

FIG. 29.—The Synthesis of Water.

Such, then, being the nature of the chemical process before us, let me pass on to that feature of this flame which is at once the most conspicuous and the most important phase of the phenomenon, namely, the development of energy. Here, again, we have become acquainted with the important facts bearing on this question. In a previous lecture I told you that, in the burning of a pound of hydrogen, sufficient energy was developed to raise a weight of 47,888,400 pounds to the height of one foot, and these figures are included, among other data of the same kind, in the diagram still before you. (*See* page 186.) I also endeavored to impress on your minds the magnitude of this energy by showing that, with a hydrogen-flame, a temperature can be obtained at which steel burns like tinder. In that experiment, however, the energy was intensified to a far greater degree than in the flame we have here; for, although this flame is very hot, it is wholly inadequate to produce the effects you before witnessed. The intensity was then gained just as in our experiment with phosphorus, by burning the hydrogen in pure oxygen, instead of air; and you remember the apparatus, called the compound blow-pipe, by which this result was obtained.

The flame of the blow-pipe emits a pale-blue light, but is so slightly luminous that it can hardly be seen at any distance in this large hall, and yet, as we know, it is intensely hot. You have seen how steel deflagrates before it, and I will now show you its effect on several other metals (copper, zinc, silver, and lead). You notice that they all burn freely, and that each imparts to the flame a characteristic color, and, I may add, in passing, that spectrum analysis, which has achieved such great results during the last few years, is based on these chromatic phenomena.

But the experiments you have just seen, although so brilliant and instructive, have not yet given us much help toward the solution of the problem we proposed to investigate, viz., the conditions under which the energy of combustion is manifested in the form of light. They have, however, helped us thus far: they have shown that the light cannot depend upon the rapidity of the combustion or the temperature of the flame alone, for here we have intense energy and a very high temperature without light. Moreover, they have presented us with a phenomenon, which differs from that we witnessed at the close of the last lecture, in the very point we are investigating: phosphorus burns in oxygen with a most brilliant light; hydrogen burns in oxygen with scarcely any light.

Now, it is evident that the cause of the light must be some circumstance of the first experiment, which does not exist in this, and, by comparing the two together, we may hope to reach a definite result. At first sight, this comparison reveals only resemblances. Both processes consist in the union of combustible material with oxygen. In the one case it is the atoms of phosphorus, and in the other the atoms of hydrogen, which combine with the atoms of the oxygen gas. Otherwise the chemical change is the same in both cases, and we cannot therefore refer the light to any difference in the process. Again, in both processes a very large amount of energy is developed, but, so far as there is any difference, that difference is in favor of the hydrogen, which gives the least light. So, also, in both processes, a very high temperature is attained; but a simple calculation will show that the temperature of the hydrogen-flame is higher than that of the phosphorus-flame, and so the light cannot be an effect solely of temperature. Can it be that the difference is due to the circumstance that the combustible in one case is a solid, and in the other a gas? Here, at least, is a difference, which gives us a starting-point in our investigation. But we shall not pursue the investigation far before we find that this difference is wholly illusory. It will appear that phosphorus is a very volatile solid, and that it is wholly converted into vapor before burning; so that, in fact, we are dealing in both cases with burning gas.

In looking round for other differences we shall recognize that there is a marked difference in the products of the two processes. The product in one case is phosphoric oxide, and in the other case water. Water is volatile, and is evolved in the state of vapor. Phosphoric oxide is a highly-fixed solid, and condenses in those snow-like flakes which you saw falling in the jar at the last lecture. May it not be that the circumstance that the product in the one case is a solid, and in the other a gas, is the cause of the difference in the light? In the phosphorus-flame there are solid particles of phosphoric oxide, while in the hydrogen-flame there are no solid particles whatever. Can this be the cause of the difference? Here, at least, is another starting-point for our investigation.

An obvious mode of discovering whether there is any value in this suggestion is to introduce non-volatile solid matter into the blow-pipe flame, and observe whether the light of the flame is affected thereby. The temperature of the flame is so high that there are but few solids which are sufficiently fixed for our experiment. One, however, which is admirably adapted for our purpose, is at hand, and that is lime. In order, then, to answer the question that has been raised, let us introduce into the flame a bit of lime, or, what amounts to the same thing, allow the flame to play against a cylinder of this material. (In an instant the hall is most brilliantly illuminated.) The question is answered, and there is no plainer answer than that given by a well-considered experiment.

And here let me ask your attention to the method we have followed, because it illustrates, in the most striking manner, the method of science. When we wish to discover the cause of an effect observed in any phenomenon, we begin by varying the conditions of the phenomenon until at last we find that the effect varies, or perhaps even disappears. That is, we try a series of experiments, varying the conditions at each trial, until at last we succeed in eliminating the effect. This having been done, we next compare the conditions under which the effect appears and those under which it does not. Those conditions common to both experiments are at once eliminated, while those which are different in the two are carefully considered, and experiments are devised to test their influence on the effect until at last the cause is made evident. Thus we sought to find the cause of the light generally produced by combustion. We began by burning different combustibles until we found one which gave out little or no light. We next compared the burning of phosphorus in oxygen, which gave a very intense light, with the burning of hydrogen, which gave little or none. We found that the only important difference between the two cases was the circumstance that the phosphorus-flame contained particles of solid matter, while the hydrogen-flame contained none, and in order to test the effect of the difference, which the comparison suggested, we placed solid matter in the hydrogen-flame, when the cause of the light became evident. This method of comparing phenomena as a means of discovering the cause of effects which are prominent in one, although common to both, is frequently called differentiation, and it is one of the most valuable methods of science. If I have succeeded in giving you some idea of the method, the time we have devoted to these experiments has been well spent.

You will grant, I think, that we have now established the following points in regard to the theory of combustion: 1. That the process requires a certain degree of molecular activity, measured roughly by what we call the point of ignition. 2. That the chemical change consists simply in the union of the combustible with the oxygen of the air. 3. That these processes differ from other examples of synthesis chiefly in the circumstance that the union of the oxygen atoms with those of our ordinary combustibles is attended with an extraordinary development of energy. 4. That the amount of this energy is constant for the same combustible, and is in each case exactly proportional to the amount of fuel burnt. 5. That the intensity of the effect depends on the rapidity of the combustion, the energy usually manifesting itself as heat, but taking also the form of light when non-volatile solid particles are present.[1]

Were we to limit our regards solely to the theory of combustion, there would be no necessity of pursuing the subject further; but additional experiments may be of value by helping you to associate these principles with your previous experience. To this end I propose to ask your attention to the burning of one of the most familiar combustibles, viz., carbon in the form of charcoal, and, in order to hasten the process, we will burn the charcoal in oxygen gas instead of air. Placing, then, a few lumps of charcoal, previously ignited, in a deflagrating spoon, I will introduce them into this large jar of oxygen gas. . . . As you see, the charcoal burns more brilliantly than in air. But even in the pure gas the burning is by no means very rapid, and the reason is obvious. Since carbon, in all its

[1] In order to give a complete view of the subject, it would be necessary to show further that liquids, and even vapors, under certain conditions, may become brilliant sources of light.

forms, is non-volatile, the molecules of the charcoal cannot leave the solid lumps. They do not, therefore, go half-way to meet the oxygen-molecules, but simply receive those which are driven against the surface of the coals. Hence the process depends on the activity of the oxygen-molecules alone, and, since the number of these molecules which can reach the combustible in a given time is limited by the extent of its surface, it is evident that with these lumps of coal we cannot expect very rapid burning even in pure oxygen. If, however, our theory is correct, we should greatly increase the rapidity by breaking up the lumps, and thus increasing the surface of contact with the gas. Let us see if the result answers our expectations.

Taking, then, some finely-pulverized charcoal, already ignited (by heating the mass in an iron dish over a spirit-lamp), I will sift the red-hot powder from an iron spoon into another large jar filled with oxygen. . . . Nothing we have yet seen has exceeded the splendor of the chemical action which now results. This dazzling light is radiated by the glowing particles of charcoal, which, after they have become incandescent, retain their solid condition until the last atom of carbon is consumed, giving us another illustration of the influence of this circumstance on the light: and let me again call your attention to the great fixity of carbon which the experiment also illustrates, and you will at once recognize the importance of this quality of the elementary substance in localizing our fires, as well as limiting their intensity, and will see that the use of coal as fuel wholly depends upon it.

Turn next to the chemical change itself. This, as in the other similar processes we have studied, is an example of simple synthesis, consisting in the union of the carbon atoms with oxygen. As to the nature of the product formed, a single experiment will give you all the information you desire.

After removing the deflagrating spoon with the residue of the charcoal lumps from the first of the two jars, I will ask you to notice the fact that the atmosphere within remains as transparent as before. The eye can detect no evidence of change, yet all the charcoal that has disappeared has been taken up by this atmosphere, and, could we readily weigh the mass of gas, I could show you that the weight had been increased by the exact weight of the coal absorbed. Indeed, the density has been so greatly enhanced that I can pour the gas from one vessel to another very much as I would water. Let me pour some of it from the jar into a tall glass half filled already with lime-water. . . . It looks like child's-play; but the transfer has been made, and now, on shaking the gas and lime-water together, the liquid becomes milky.

You at once recognize the product: chalk has been formed in the lime-water, and the gas left after the burning ceased in the jar must be the same carbonic dioxide we have previously studied. We made the analysis of this aëriform substance in a previous lecture, and we have now made the synthesis. See how simply we express the reaction:

$$\underset{\text{Coal.}}{C} + \underset{\text{Oxygen Gas.}}{O{=}O} = \underset{\text{Carbonic Dioxide.}}{CO_2.}$$

A fact is indicated by this reaction, which we must not overlook. The volume of the carbonic dioxide (CO_2) obtained is exactly equal to the volume of the oxygen gas (O=O) employed. In this experiment we used a jarful of oxygen and we obtained a jarful of carbonic dioxide. The material of the burnt charcoal is taken up into the gas atom by atom, actually absorbed by it as a sponge absorbs water. Every molecule of oxygen which strikes against the charcoal flies off with an atom of carbon, forming with it the molecule of carbonic dioxide which, of course, occupies the same space as the previous molecule of oxygen gas. Hence it is that the vast amount of carbon which is being constantly absorbed by the atmosphere, as it passes through our grates and furnaces, does not alter its volume. Would that I might impress this remarkable fact on your imagination! Consider how much coal is being burnt every day in a city like this—hundreds and hundreds of tons! Conceive of what a mass it would make, more than filling this large hall from floor to ceiling, and yet in our city alone this enormous black mass is in twenty-four hours absorbed by the transparent air, picked up and carried away bodily, atom by atom, by the oxygen-molecules.

Turn now to the energy developed in this process. Our diagram indicates that the amount of energy developed by the burning of a pound of coal is very much less than that obtained with a pound of hydrogen. But then it must be remembered how attenuated hydrogen gas is; if, instead of comparing equal weights, we compare equal volumes, we shall find that the difference is vastly in favor of carbon.

Most of the combustible materials, however, which we use as fuel, consist of both hydrogen and carbon; but the phenomena we have studied in the burning of the elementary substances reappear with these familiar combustibles, and, in regard to them, there are only a few special points to be noticed. On many of these substances, such as naphtha, paraffine, stearine, wax, oil, and the like, the effect of the heat is to generate illuminating gas, which is the source of most of our artificial light. In our cities and large towns the gas is made for us by a special process, but it must be remembered that every lamp and candle is a small gas-factory. Flame is always burning gas, and the gas which we burn in our lamps and candles is very similar to that supplied by the Boston Gas Company: the only difference is that the gas, instead of being made from bituminous coal, is made from petroleum or wax, and, instead of being made at the "North End" and distributed through pipes to distant burners, is burnt as fast as it is made. The heat generated by the burning gas is so great that it volatilizes the oil or wax fast enough to supply the flame, and then the mechanism of the wick comes into play to keep the parts of these natural gas machines in perfect running order. Indeed, a common candle, simple as it appears to be, is a most wonderful apparatus, and I should be glad to occupy the whole hour in explaining the adaptation of its parts; but I have only time for a few illustrations, which show that in these luminous flames, as in the other cases of combustion we have studied, the light comes from incandescent solid particles.

Of the two constituents of the combustible gas which forms the flame, hydrogen is the most combustible, and under ordinary conditions is the first to burn, setting free, for a moment, the accompanying carbon in the form of a fine soot which fills the light-giving cone. This dust is at once intensely heated, and each glowing particle becomes a centre of radiation, throwing out its luminous pulsations in every direction. The sparks last, however, but an instant, for the next moment the charcoal is itself consumed by the fierce oxygen, now aroused to full activity, and only a transparent gas rises from the flame. But the same process continues; other particles succeed, which become ignited in their turn, and hence, although the sparks are evanescent, the light is continuous.

I might illustrate this theory by the familiar fact that soot is at once emitted from all these luminous flames, whenever the draft becomes so far interrupted that it does not supply sufficient oxygen to burn completely the carbon particles; but a still more striking illustration is furnished by the simple contrivance we employ in the laboratory for preventing the deposition of this soot on the heating surfaces of our chemical vessels. We use for this purpose a gas-burner invented by Prof. Bunsen, of Heidelberg, and known by his name, in which air is mixed with the hydrocarbon gas before it is burnt. But this air, while it prevents the formation of soot, at the same time destroys the illuminating power of the flame. The molecules of the hydrocarbon gas being now in near proximity to the molecules of oxygen required for complete combustion, the difference of affinity of oxygen for the carbon and hydrogen atoms does not come into play. There is enough oxygen for all, and the result is that no carbon-particles are set free in the flame. We have no soot, and therefore no light.

In this Bunsen lamp the size of the apertures, by which the air enters at the base of the burner, may be regulated by a valve, and you notice that on closing this valve the flame at once becomes luminous. Open it again so that the gas shall mix with air before burning, and the energy no longer takes the form of light. See, nevertheless, how brightly the flame ignites this coil of platinum wire, showing that there is no want of energy, only it now appears wholly as heat.

The flame of a wood or soft-coal fire is also a gas-flame. The first effect of heat on these bodies is to generate illuminating gas, and to this circumstance, as in the case of the candle, the flame is due, but after a while all the hydrogen is driven off, and we have then, in the glowing embers, the flameless combustion of carbon.

The chemical change which takes place in the burning of hydrocarbon fuels is in no way affected by the circumstance that the hydrogen and carbon are in chemical union. All the hydrogen-atoms burn to water, and all the carbon-atoms to carbonic dioxide, and these products can be detected in the smoke of every flame; indeed, with a few unimportant exceptions, they are the sole products of the combustion.

Take, for example, this candle-flame. On holding over it a cold bell-glass the glass soon becomes bedewed, and, before long, drops of water begin to trickle down the sides; and now, on inverting the bell, and shaking up in it some lime-water, the milky appearance, which the clear solution immediately assumes, indicates the presence of carbonic dioxide.

Of course, all the material of the candle passes into these colorless and insensible aëriform products which mingle with the atmosphere, and this absorption of combustible material into the atmosphere, this melting of firm, solid masses of coal and wood into thin air, has such an appearance of annihilation that it requires all the power of the reason, aided by experiment, to correct the false impression of the senses. Yet nothing is easier than to show that the smoke, colorless and insensible as it is, weighs more than the material burnt, and, although the experiment must be familiar to many of my audience, I will repeat it, because it may aid some to clearer views of this all-important subject.

Let me call your attention, then, to this candle which, in a candlestick of peculiar construction, is hanging equipoised from one end of the beam of this balance (Fig. 30). You know that both aqueous vapor and carbonic dioxide are eagerly absorbed by caustic soda, and this apparatus is so arranged that the smoke of the candle is sucked through two glass tubes filled with this absorbent material. You notice that my balance is in equilibrium, and I will now light the candle under its tin chimney. The products of the combustion rise to the top of the chimney, which is closed excepting two small apertures, through which the smoke is sucked into the glass tubes containing the caustic soda. Now you must picture to yourselves the molecules of oxygen of our atmosphere rushing in on this candle-flame from every side, each one seizing its atom of carbon, or its four atoms of hydrogen, as the case may be. You must, then, follow the molecules of carbonic dioxide and water thus formed, as they are caught up by the current of air—which our aspirator draws through the apparatus—and hurried into the glass tubes, where they are seized upon and held fast by the caustic soda. All the smoke of the candle being thus retained, it is evident that, if the process is as I have described it, we should expect that the apparatus would increase in weight as the candle burns, while, on the other hand, were any part of the material lost, there would be a corresponding diminution in weight. And we not only find that the weight increases, as the bal-

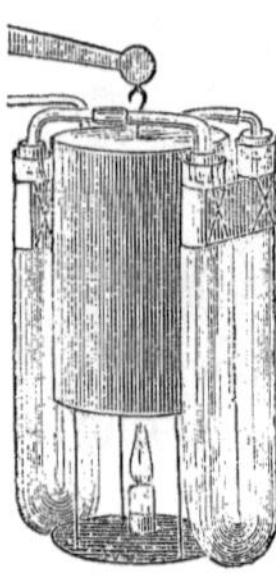

Fig. 30.

ance shows, but that the increase is exactly equal to the amount of oxygen consumed. Not only none of the material of the candle escapes from the apparatus, but a portion of the oxygen of the air is also retained, and that causes the increase of weight.

In connection with this experiment, I must not fail to call your attention to the circumstance that the products of this combustion are as harmless as they are imperceptible to the senses. Remember that thousands of tons of carbonic dioxide and aqueous vapor are discharged into the air of this city in a single day. Remember, also, what a howl of remonstrance goes up if, from some manufactory, a few pounds of similar but noisome products escape, and you cannot fail to recognize the importance of this fact in the economy of Nature. Add to this what you already know, that the smoke of our fires and the exhalations of our lungs is the food of the plant—that the whole vegetable world is constantly absorbing carbonic dioxide, and giving back the oxygen to the atmosphere while storing up the regenerated carbon in its tissues, and you will be still further impressed by the wonderful revelations we are studying.

Nor must we, in this connection, fail to notice again the enormous amount of energy which the burning of our common forms of fuel liberates. The table is still before you which shows how great is the amount of energy which can be obtained by the burning of a single pound either of hydrogen gas or of charcoal, and the relations of these elementary substances in this respect are not in the least altered by their association in common wood or coal. In round numbers, it may be said that a cubic foot of cannel coal contains sufficient energy, if wholly utilized, to raise a weight of 3,269 tons one hundred feet, or 732,000,000 pounds one foot. I said, if wholly utilized, for, although we are able to make use of the whole energy in the form of heat, we have not yet succeeded in applying more than about one-twentieth of it to mechanical work. But still the energy exists stored up for use in every foot of wood or coal, and is ready to be set free when the fuel is burnt.

When standing before a grand conflagration, witnessing the display of mighty energies there in action, and seeing the elements rushing into combination with a force which no human energy can withstand, does it seem as if any power could undo that work of destruction, and rebuild those beams and rafters which are melting into air? Yet, in a few years thay will be rebuilt. This mighty force will be overcome; not, however, as we might expect, amid the convulsions of Nature or the clashing of the elements, but silently in a delicate leaf waving in the sunshine. As I have already explained, the sun's rays are the Ithuriel wand, which exerts the mighty power, and under the direction of that unerring Architect, whom all true science recognizes, the woody structure will be rebuilt, and fresh energy stored away to be used or wasted in some future conflagration.

My friends, this is no theory, but sober, well-established fact. How the energy comes and how it is stored away, we attempt to explain by our theories. Let these pass. They may be true, they may be mere fancies; but, that the energy comes, that it is stored away, and that it does reappear, are as much facts as any phenomena which the sun's rays illuminate. I know of no facts in the whole realm of Nature more wonderful than these, and I return to them in the annual round of my instruction with increasing wonder and admiration, amazed at the apparent inefficiency of the means, and the stupendous magnitude of the result. In another course of lectures in this place I endeavored to show what weighty evidence these facts give in support of the argument that all the details have been arranged by an intelligent Designer.[1] The plan of this course does not give me time to do more than allude to this point, and I only refer to it here to ask for the argument your own careful consideration.

There is still another point, in connection with this subject, to which also I can only barely allude. The crust of our globe consists almost wholly of burnt material. Our granite, sandstone, and limestone rocks, are the cinders of the great primeval fire, and the atmosphere of oxygen the residue left after the general conflagration—left because there was nothing more to burn. Whatever of combustible material, wood, coal, or metal, now exists on the surface of the earth, has been recovered from the wreck of the first conflagration by the action of the sun's rays. One-half of all known material consists of oxygen, and, on the surface of the globe, combination with oxygen is the only state of rest. In the process of vegetable growth, the sun's rays have the power of freeing from this combination hydrogen and carbon atoms, and from these are formed the numberless substances of which both the vegetable and animal organisms consist. From the material of these organisms we make charcoal, and Nature makes her coal-beds, and supplies her petroleum-wells. Moreover, with these same materials, man has been able to separate the useful metals from their ores, and, by the aid

[1] "Religion and Chemistry; or, Proofs of God's Plan in the Atmosphere and its Elements," ten lectures by Josiah P. Cooke, Jr, published by Charles Scribner. New York, 1864.

of various chemical processes, to isolate the other elementary substances from their native compounds; but the efficiency of all these processes depends on employing the energy which the sun's rays impart to the carbon and hydrogen atoms to do work. A careful analysis of the conditions will show that it is just as truly the sun's energy which parts the iron from its combination in the ore, as it is solar power which parts the carbon from the carbonic dioxide in the leaf. We have here, however, but a single example of a general truth. All terrestrial energy comes from the sun, and every manifestation of power on the earth can be traced directly back to his energizing and life-giving rays. The force with which oxygen tends to unite with the other elements may be regarded as a spring, which the sun's rays have the power to bend. In bending this spring they do a certain amount of work, and, when, in the process of combustion, the spring flies back, the energy reappears. Moreover, the instability of all organized forms is but a phase of the same action, and the various processes of decay, with the accompanying phenomenon of death, are simply the recoiling of the same bent spring. Amid all these varied phenomena, the one element which reappears in all, and frequently wholly engrosses our attention, is energy; and, if I have succeeded in fixing your attention on this point, my great object in this lecture has been gained. In the early part of this course, I stated that all modern chemistry rests on the great truth that MATTER IS INDESTRUCTIBLE, AND IS MEASURED BY WEIGHT. This evening we have seen glimpses of another great central truth, which, although more recently discovered, is not less far-reaching or important, namely, ENERGY IS INDESTRUCTIBLE, AND IS MEASURED BY WORK. Add to these two a third, namely—INTELLIGENCE IS INDESTRUCTIBLE, AND IS MEASURED BY ADAPTATION—and you have, as it seems to me, the three great manifestations of Nature: MATTER, ENERGY, and INTELLIGENCE. These great truths explain and supplement each other. Give to each its due weight in your philosophy, and you will avoid the extremes of idealism on the one side, and of materialism on the other.

NOTE.—The doctrine that energy is indestructible is known in physics as the *conservation of energy*, and it would give greater prominence to the corresponding doctrine—that matter is indestructible—if it were called the *conservation of mass*. The recognition of this last truth by Lavoisier has already been indicated, and its important influence on the history of chemical philosophy has been discussed; but the student is not likely to appreciate its full significance unless he dwells upon it, and a comprehensive view of the subject will probably be best gained by bringing the chemical changes, in which the truth can only be verified by careful investigation, into comparison with those familiar mechanical processes in which it is perfectly obvious. That there is no annihilation of material in the conversion of water into oxygen and hydrogen gases is a phase of the same truth, which we accept as self-evident in the common mechanical operations of the arts. Just as gold coins contain the metal from which they were struck, so the oxygen and hydrogen gases contain the material of the water from which they were made, and, if the student fully grasps the truth which this statement involves he will, in the first place, perceive that *mass* may be an attribute of matter underlying those accidents in which substances differ; and, in the second place, he will see that the assumption of immutable atoms is an obvious explanation of the conservation of mass in Nature.

LECTURE X.

GUNPOWDER AND NITRO-GLYCERINE.

THERE is one further point in connection with the theory of combustion to which I wish to call your attention, at the outset of my lecture this evening. In the only cases of burning we have studied, the combustible unites with the oxygen of the atmosphere. It is possible, however, to have combustion without atmospheric air, the combustible obtaining the required oxygen from some associated substance. There are several substances in which a large amount of oxygen is so loosely combined, or, in other words, in which the oxygen-atoms are held in combination by such a feeble force, that they will furnish oxygen to the combustible as readily as the atmosphere, and in a vastly more concentrated form. Two of these substances are well known, nitre (potassic nitrate) and chlorate of potash (potassic chlorate). One ounce of this last salt—the quantity in this small crucible—contains enough oxygen to fill a large jar (1.7 gallon), and by simply heating the salt we should obtain that amount of oxygen gas. We have provided also one-third of an ounce of pulverized sugar, and we will now mix the two powders thoroughly together. Consider the conditions in this mixture: The sugar is a combustible substance, and every particle of this combustible is in contact with, or, I should rather say, in close proximity to, grains of chlorate of potassa, which contain sufficient oxygen to burn the whole. All is now quiescent, because both materials, being in the solid condition, their molecules are, as it were, imprisoned, and a certain degree of molecular activity is required to produce chemical change. This molecular activity we can readily excite by heat, but a more convenient, although less intelligible way, is to touch the mixture with a drop of sulphuric acid.

Here we have not merely a pretty firework, but an experiment which illustrates a very important phase of the phenomena of combustion, and one of immense practical value. I have chosen this particular example because you are familiar with both of the materials employed. You have seen that sugar contains a large amount of combustible carbon. You also know that potassic chlorate contains a large volume of oxygen, which can readily be driven off by heat; for you have seen me make oxygen from this very salt. You can, therefore, fully appreciate the conditions we had in our crucible at the beginning of the experiment, namely, a combustible with the oxygen required to burn it in close proximity. You will be prepared, then, to understand—1. That the burning we have just witnessed does not differ from ordinary burning, except in the single point I have mentioned; that the combustible derives its oxygen from potassic chlorate, instead of from the air; and, 2. that it is possible to inclose in a confined space, as a gun-barrel or a bomb, all the conditions of combustion. In a word this experiment illustrates the simple theory of gunpowder.

What, then, is gunpowder? Essentially a mixture

of two substances—saltpetre and charcoal, with merely a small amount of sulphur added to facilitate the kindling of the charcoal. In the manufacture of this explosive agent, as is well known, the materials are first reduced to a very fine powder, and then intimately mixed together. Afterward, by great pressure, the mass is compacted to a firm, hard cake, which is subsequently broken up into grains of different sizes, adapted to various uses. Here we have some samples of these grains, varying from the size of a walnut to that of a millet-seed. These black grains, although they appear so homogeneous, are, in fact, a very intimate mixture of a combustible material (charcoal and a little sulphur) with a substance rich in oxygen (saltpetre), and, when we ignite the powder, the charcoal burns at the expense of the oxygen of the saltpetre. Two parallel experiments will make the whole matter clear.

In this jar we have about one gallon (100 grains) of pure oxygen, enough to combine with 37½ grains of charcoal. This quantity of charcoal we will place in a copper spoon, and, having ignited the coal, we will plunge it into the jar of oxygen. We have at once a brilliant combustion, and a repetition of the experiment which you witnessed at the last lecture. We then learned that the process consists in the union of the oxygen with the carbon, and that each molecule of oxygen gas actually picks up an atom of carbon to form a molecule of carbonic dioxide. There are, therefore, just as many molecules in the jar at the close of the experiment as at the first, only they now consist of three atoms, instead of two; O=O has become O=C=O.

In the second jar is a cup containing a small quantity of gunpowder, and so arranged that the powder can be exploded by a voltaic battery. As the oxygen-atoms required for the burning are lying in the cup side by side with the charcoal, we do not need the air in our experiment. Accordingly, we have connected the jar with an air-pump, so that we can exhaust the air. . . . The gauge of the pump now indicates that the greater part of the air has been removed. Notice further that, when we readmit a little air, the mercury column falls, and thus, as you see, this gauge will tell us when any gas enters the jar. . . . Having again completed the exhaustion, let us fire the powder. . . . The powder has disappeared; but the gauge indicates that a large volume of gas has been formed.

A simple test will now show that the aëriform products in the two last experiments are identical. Here are two glasses, each filled with lime-water. To one we will add some of the gas from the first jar, pouring it in upon the lime-water, and to the other we will add some of the gas from the gunpowder, by pouring as before. On shaking the gas and liquid together, we obtain in both cases the familiar milky turbidness which indicates the presence of carbonic dioxide. It is true that the carbonic dioxide from the gunpowder is not quite so pure as that found in the other jar, but this is an unessential matter.

Having seen that gunpowder, burnt in a vacuum, is quietly resolved into gas, we will next take an equal amount of powder and inclose it in a pasteboard case, which we call a cartridge, using the same arrangement for firing the powder as before. We make the connection, and off it goes! . . . There can be no occasion, I think, to seek far for the cause of the explosion. The chemical process must have been identical with that in our jar; but, while in the jar there was room for all the gas-molecules formed in the burning, the small volume of the cartridge could not hold them, and they burst out, tearing away the paper walls in their course. The gas evolved would occupy, at the ordinary pressure of the air, about three hundred times the volume of the powder used, and, if confined in the space previously filled with the powder, would exert a pressure equal to about $300 \times 14 = 4{,}200$ lbs., or two tons, on a square-inch. The pressure obtained is really far greater than this, on account of the heat developed by the combustion. Moreover, as the powder burns rapidly, this pressure is suddenly applied, and has all the effect of an immensely heavy blow, which no strength of materials is sufficient to withstand. Of course, any chamber in which the powder is confined gives way at the weakest point. In the chamber of a gun the ball usually yields before the breech, and is hurled with violence from the mouth of the piece; but fearful accidents not unfrequently occur when, for any reason, the ball has been too tightly wedged, or when the metal of the breech is too weak.

You all know that a large amount of gas condensed into a small chamber must exert great pressure, and therefore you will undoubtedly regard the explanation I have given of the force exerted by gunpowder as satisfactory and sufficient. But, although this is the usual way of presenting the phenomena, I am anxious that you should view them in the light of our modern molecular theory, which gives to the imagination a far more vivid picture of the manner in which the power acts.

Begin with the black grains as they lie in the chamber of the gun behind the ball. You must remember that all the ingredients of the powder are in a solid condition, and picture to your imagination the molecules as held in their places by those forces which I attempted to make evident to you in a former lecture, incapable of any motion except a slight oscillation about the centres of force. The gun is now fired, and the powder burns. We need consider but two of the immediate consequences: first, there is a large volume of gas formed; and, secondly, there is a very great amount of energy developed. Picture to yourselves, now, an immense number of gas-molecules suddenly set free in the chamber of the gun, and animated with all the velocity which great energy is capable of imparting. See these molecules rushing against the ball with their whole might, and, when at last it starts, imparting to the projectile their moving power, until it acquires the fearful velocity with which it rushes from the mouth of the gun. The molecules impart their motion to the ball, just as one billiard-ball imparts motion to another. The effect is due to the accumulation of small impulses; for, although the power imparted by a single molecule may be as nothing, the accumulated effect of millions on millions of these impulses becomes immense.

Within a few years our community have become familiar with the name and terrible effects of a new explosive agent, called nitro-glycerine, and I feel sure that you will be glad to be made acquainted with the remarkable qualities and relations of this truly wonderful substance. Every one knows that clear, oily, and sweet-tasting liquid called glycerine, and probably most of you have eaten it for honey. But it has a great many valuable uses, which may reconcile you to its abuse for adulterating honey, and it is obtained in large quantities as a secondary product of the manufacture of soap and candles from our common fats. Now, nitro-glycerine bears the same relation to glycerine that saltpetre bears to caustic potash. Common saltpetre, which is the oxygenated ingredient of gunpowder, is called in chemistry potassic nitrate, and, although the commercial supply comes wholly from natural sources, it can easily be made by the action of nitric acid on caustic potash. My assistant will pour some nitric acid into a solution of caustic potash, and you will soon see crystals of saltpetre appear, shooting out from the sides of the dish, whose image we have projected on the screen. In a similar way we can prepare nitro-glycerine by pouring glycerine in a fine stream into very strong nitric acid, rendered more active by being mixed with sulphuric acid—oil of vitriol.

We could easily make the experiment, but you could see nothing. There is no apparent change, and it is a remarkable fact that, when pure, nitro-glycerine resembles, externally, very closely glycerine itself, and, like it, is a colorless, oily fluid—the reddish-yellow color of the commercial article being due to impurities. As soon as the chemical change is ended, the nitro-glycerine must be very carefully washed with water, until all adhering acid has been removed. The material thus obtained has most singular qualities, and not the least unexpected of these is its stability under ordinary conditions. After the terrible accidents that have happened, it would, perhaps, be rash to say that it did not readily explode ; but I can assure you that it is not an easy matter to explode pure nitro-glycerine. It is not nearly so explosive as gunpowder, and I am told that the flame of an ordinary match can be quenched in it without danger, although I confess that I should be unwilling to try the experiment. Still, there can be no doubt that, under ordinary circumstances, a small flame will not ignite it. My knowledge of the matter is derived from Professor Hill, of the Torpedo Station at Newport, who has studied very carefully the preparation and application of the material. He is of opinion that most of the accidents which have given to nitro-glycerine such an unfortunate notoriety have been caused by the use of an impure article, and that proper care in its preparation would greatly lessen the danger attending its use. Nitro-glycerine is usually exploded, not by the direct application of heat, but by a sudden and violent concussion, which is obtained by firing in contact with it a fuse of some fulminating powder. The effects of this explosion are as peculiar as the method by which it is obtained, and I can best illustrate the subject by describing an experiment with nitro-glycerine which I witnessed myself at the Torpedo Station a few months since.

It is so inconvenient to handle liquid nitro-glycerine that it is now usual to mix it with some inert and impalpable powder, and the names dualine and dynamite have been given to different mixtures of this kind; but in both of these the powder merely acts as a sponge. In the experiment referred to, a canister holding less than a pound of dynamite, and only a few ounces of nitro-glycerine, was placed on the top of a large bowlder-rock, weighing two or three tons. In order that you may fully appreciate the conditions, I repeat that this tin case was simply laid on the top of the bowlder, and not confined in any way. The nitro-glycerine was then exploded by an appropriate fuse fired from a distance by electricity. The report was not louder than from a heavy gun, but the rock on which the canister lay was broken into a thousand fragments.

This experiment strikingly illustrates the peculiar action of nitro-glycerine. In using gunpowder for blasting it is necessary to confine it, by what is called tamping, in the hole prepared for it in the rock. Not so with nitro-glycerine. This, though it may be put up in small tin cartridges for convenience, is placed in the drill-holes without tamping of any kind. Sometimes the liquid itself has been poured into the hole, and then a little water poured on the top is the only means used to confine it. As an agent for blasting, nitro-glycerine is so vastly superior to gunpowder that it must be regarded as one of the most valuable discoveries of our age. Already it is enabling men to open tracks for their iron roads through mountain-barriers which, a few years ago, it would have been thought impracticable to pierce, and, although its introduction has been attended with such terrible accidents, those best acquainted with the material believe that, with proper care in its manufacture, and proper precautions in its use, it can be made as safe as or even safer than gunpowder, and the Government can do no better service toward developing the resources of the country than by carrying forward the experiments it has instituted at the Torpedo Station at Newport, until all the conditions required for the safe manufacture and use of this valuable agent are known, and, when this result is reached, imposing on the manufacturers, dealers, and carriers, such restrictions as the public safety requires. Of course, we cannot expect, thus, to prevent all accidents. Great power in the hands of ignorant or careless men implies great danger. Sleepless vigilance is the condition under which we wield all the great powers of modern civilization, and we cannot expect that the power of nitro-glycerine will be any exception to the general rule.[1]

But, while nitro-glycerine has such great rending power, it has no value whatever as a projectile agent. Exploded in the chamber of a gun, it would burst the breech before it started the ball. Indeed, there is a great popular misapprehension in regard to the limit of the projectile power of gunpowder, and inventors are constantly looking for more powerful projectile agents as the means of obtaining increased effects. But a study of the mechanical conditions of projection will show not only that gunpowder is most admirably adapted to this use, but also that its capabilities far exceed the strength of any known material, and the student will soon be convinced that what is wanted is not stronger powder, but stronger guns. I do not mean to say that we cannot conceive of a better powder than that now in use, but merely that its shortcoming is not want of strength.

Having described the properties of nitro-glycerine, the question at once arises, "Can these singular properties be explained?" In order to answer this question I shall next ask your attention to the theory of its action, and I think you will find that our modern chemistry is able to give a very intelligible account of the phenomena we have described. I will begin by saying that the chemical action in the explosion of nitro-glycerine is very similar to that in the burning of gunpowder. In both cases we have the same two results: 1. The production of a large volume of gas; 2. The

[1] The recent improvements in the manufacture of gun-cotton, and the discovery that, even when too wet to burn, it can be exploded by concussion if the fuse is sufficiently powerful, promise to furnish an explosive agent nearly equal to nitro-glycerine in strength, and free from all ordinary risks.

liberation of a large amount of energy which gives to the confined gas-molecules an immense moving power. Moreover, essentially the same aëriform products are formed in the two cases, and in both the process consists, for the most part, in the union of carbon and hydrogen atoms with oxygen. But, while in the gunpowder the carbon and oxygen atoms are in different molecules, although lying side by side in the same grains, in the nitro-glycerine they are in different parts of the same molecule. And here comes our first glimpse of the most recondite chemical principle the science has yet attained, one which I have been aiming to reach throughout this whole course of lectures, and one which it will be my object in the three remaining lectures clearly to set before you. I can, as yet, only state the principle as a theorem to be proved; but, if I can succeed in making this difficult subject clear, I feel confident that you will regard the proof as satisfactory. The principle is this:

Every molecule has a definite structure. It not only consists of a definite kind and a definite number of atoms, but these atoms are arranged or grouped together in a definite order, and it is the great object of modern chemistry to discover what that grouping is. Almost all the great chemists of the world are, at this moment, engaged in investigating this very problem, and, what is more, they have succeeded, in many cases, in solving it, and we have reached as much certainty in regard to the grouping of the atoms in the molecules of a very large number of substances, as we have in regard to any phenomena so wholly super-sensible. For example, we feel well assured that we know how the atoms are grouped in the molecule of nitro-glycerine, and the diagram before you represents in

```
    H     O
    |     ||
H - C - O - N = O
    |
H - C - O - N < O
                O
    |
H - C - O - N = O
    |     ||
    H     O
```

Order of Atoms in the Molecule of Nitro-glycerine.

```
O       H   H   H       O
||      |   |   |       ||
N - O - C - C - C - O - N
||      |   |   |       ||
O       H   O   H       O
            |
          O = N = O
```

Same Order, but different Form of Symbol.

our rude way the result we have reached. The letters signify single atoms, and the lines between the letters merely show how the atoms are severally united. Begin with the three atoms of carbon, which are united together, say, by a certain force, which the lines denote. To these are directly united five atoms of hydrogen, and then to each of the carbon-atoms is also bound the atomic group $-O-N{<}^{O}_{O}$, the four atoms of the group having a definite arrangement among themselves. There is no virtue in the mere form of the arrangement of the letters on the diagram. It is perfectly possible that the atoms may be arranged so as to form regular geometrical figures, such as some theorists have amused themselves in constructing; but we do not pretend to have any accurate knowledge on this point. All we affirm is, that the atoms are united, one with another, in the order I have indicated, and the second diagram, in which the several atoms are united as before, although the form of the arrangement is different, means, to the chemist, precisely the same thing as the first.

Now, as I said, I present to you this diagram of the constitution of a molecule of nitro-glycerine simply as a theorem to be proved. As it hangs before you, I have no doubt that it will shake your faith in the credibility of the scientific investigators who bring forward this as the sober conclusion at which they have arrived. Indeed, when I first saw these attempts to represent the grouping of atoms, they appeared to me to be the vagaries of a diseased scientific imagination; for, remember, this molecule, whose structure is here portrayed, cannot be larger than the $\frac{1}{25,000,000}$ of an inch. But, as the evidence pressed upon me, I reluctantly examined it. Finding that it could not be gainsaid, I was forced to accept the conclusion, and soon I found myself busy at the same work. Now, I only ask you to accept this diagram as a theorem to be proved, and, assuming it for the time to represent, although very rudely, a real truth, see how fully it explains the properties of nitro-glycerine. Indeed, the facts already before us furnish the strongest evidence possible of the general truth of the principle I have asked you to assume; for, if you accept the principles I have previously endeavored to establish, and once admit that there are such things as molecules and atoms, the properties of nitro-glycerine will force you to admit that its molecules have a definite structure. See how the case stands.

Nitro-glycerine has been analyzed, and, unless the principles of our modern chemistry are all wrong, its molecules have the composition indicated by the symbol $C_3H_5N_3O_9$. Note that there are already in the molecule nine atoms of oxygen, more than enough to satisfy all the atoms, both of carbon and of hydrogen. When carbon burns, C_3 only takes O_6, H_5 only $O_{2\frac{1}{2}}$, and why is not the affinity of these atoms for oxygen satisfied already? The only answer that can be suggested is, because the oxygen-atoms, although parts of the same molecule, are not in combination with the carbon or hydrogen atoms in those molecules; and what is this but an admission that the molecules have a definite structure by which these atoms are kept apart?

In the next place, admitting that the structure is that represented above, you see how the atoms are kept apart. Three of the oxygen-atoms form the links, as it were, between the carbon and nitrogen atoms, and the rest of the oxygen-atoms are united with the nitrogen-atoms, and not with those of either carbon or hydrogen. Now, when the substance explodes, what takes place is simply this: The oxygen-atoms at one end of the molecule rush for the atoms of carbon and hydrogen at the other end, and the molecule is broken up, as our next diagram indicates; only, as there are not enough atoms to form even mole-

```
    H       O
    |       ‖
H - C - O - N = O      H - O - H      O = C = O
    |        ⁄O
H - C - O - N         H - O - H      O = C = O      N ≡ N
    |        ⁄O
H - C - O - N = O      H - O - H      O = C = O      N ≡ N
    |       ‖           Water.        Carbonic      Nitrogen
    H       O                         Dioxide.        Gas.
  Nitro-glycerine.
```

cules, we must consider that one atom of hydrogen and one of nitrogen are borrowed from the fragments of a neighboring molecule, broken up at the same time. You see, therefore, that the chemical action is very nearly the same as in the burning of gunpowder, the difference being that, while in the powder the carbon and oxygen atoms belong to different molecules, in nitro-glycerine they belong to the same molecule. In both cases the carbon burns, but in the nitro-glycerine the combustion is within the molecule. This difference, however, which the theory indicates, is one of great importance, and shows itself in the effects of the explosion.

In gunpowder the grains of charcoal and nitre, although very small, have a sensible magnitude, and consist each of many thousand if not of many million molecules. The chemical union of the oxygen of the nitre with the carbon-atoms of the charcoal can take place only on the surface of charcoal-grains; the first layer of molecules must be consumed before the second can be reached, and so on. Hence the process, although very rapid, must take a sensible time. In the nitro-glycerine, on the other hand, the two sets of atoms, so far from being in different grains, are in one and the same molecule, and the internal combustion is essentially instantaneous. Now, this element of time will explain a great part of the difference in the effect of the two explosions, but a part is also due to the fact that nitro-glycerine yields fully nine hundred times its volume of gas, while with gunpowder the volume is only about three hundred times that of the solid grains. There is a further difference in favor of the nitro-glycerine in the amount of energy liberated, but this we will leave out of account, although it is worthy of notice that energy may be developed by internal molecular combustion as well as in the ordinary processes of burning.

The conditions, then, are these: With gunpowder we have a volume of gas, which would normally occupy a space three hundred times as great as the grains used, liberated rapidly, but still in a perceptible interval. With nitro-glycerine a volume of gas, nine hundred times that of the liquid used, is set free, all but instantaneously. Now, in order to appreciate the difference of effect which would follow this difference of condition, you must remember that all our experiments are made in air, and that this air presses with an enormous weight on every surface. If a volume of gas is suddenly liberated, it must lift this whole weight, which, therefore, acts as so much tamping material. This weight, moreover, cannot be lifted without the expenditure of a large amount of work. Let us make a rough estimate of the amount in the case of nitro-glycerine. We will assume that in the experiment at Newport the quantity exploded yielded a cubic yard of gas. Had the air given way, instead of the rock, the liberation of this volume of gas must have lifted the pressure on one square yard (about nine tons) one yard high, an amount of work which, using these large units, we will call nine yard-tons or about 60,000 foot-pounds. Moreover, this work must have been done during the excessively brief duration of the explosion, and, it being less work to split the rock, it was the rock that yielded, and not the atmosphere. Compare, now, the case of gunpowder. The same weight of powder would yield only about one-third of the volume of gas, and would, therefore, raise the same weight to only one-third of the height; doing, therefore, but one-third of the amount of work, say 20,000 foot-pounds. Moreover, the duration of the explosion being at least one hundred times longer than before, the work to be done in lifting the atmosphere during the same exceedingly short interval would be only $\frac{1}{100}$ of 20,000 foot-pounds, or 200 foot-pounds, and, under these circumstances, you can conceive that it might be easier to lift the air than to break the rock.

If there are some who have not followed me through this simple calculation, they may, perhaps, be able to reach clear views upon the subject by looking at the phenomena in a somewhat different way. It can readily be seen that the sudden development of this large volume of gas, which becomes at once a part of the atmosphere, would be equivalent to a blow by the atmosphere against the rock; or, what would be a more accurate representation of the phenomenon, since the air is the larger mass, and acts as the anvil, a blow by the rock against the air. It may seem very singular that our atmosphere can act as an anvil, against which a rock can be split, and yet it is so, and, if the blow has velocity enough, the atmosphere presents as effective a resistance as would a granite ledge. The following consideration will, I think, convince you that this is the case: I have here a light wooden surface, say, one yard square; the pressure of the air against the surface is equal, as I just stated, to about nine tons; but the air presses equally on both sides, and the molecules have such great mobility that, when we move the surface slowly, they readily give way, and we encounter but little resistance. If, however, we push it rapidly forward, the resistance greatly increases, for the air-molecules must have time to change their position, and we encounter them in their passage. If, now, we increase the velocity of the motion to the highest speed ever attained by a locomotive—say, one and one-fifth mile per minute—we should encounter still more particles, and find a resistance which no human muscle could overcome. Increase that velocity ten times, to twelve miles a minute, the velocity of sound, and the air would oppose such a resistance that our wooden board would be shivered into splinters. Multiply again the velocity ten times, and not even a plate of boiler-iron could withstand the resistance. Multiply the velocity once more by ten, and we should reach the velocity of the earth in its orbit, about 1,200 miles a minute, and, to a body moving with this velocity, the comparatively dense air at the surface of the earth would present an almost impenetrable barrier, against which the firmest rocks might be broken to fragments. Indeed, this effect has been several times seen, when meteoric masses, moving with these planetary velocities, penetrate our atmosphere. The explosions which have been witnessed are simply the effect of the concussion against the aëriform anvil at a point where the atmosphere is far less dense than it is here. So, in the case of the nitro-glycerine, the rock strikes the atmosphere with such a velocity that it has the effect of a solid mass, and the rock is shivered by the blow.

In concluding my illustrations of the theory of combustion, a few words in regard to its history will not be out of place. We owe this theory to the great French chemist Lavoisier, who was murdered by the French communists during the reign of terror which accompanied the first French Revolution. The theory came almost perfect from his hands, and caused a revolution in the science of chemistry. Some would even date the beginning of scientific chemistry at this epoch.

In this connection, there is a recent incident which amusingly illustrates, not only the importance of theory, but also the not unfrequent contrast between the theorizing and investigating mind—the scientific poet and the scientific philosopher. About three years since, Professor Wurtz, of Paris, to whom modern chemistry owes as much as to any individual man, an Alsatian by birth, a German by descent and in many of his traits of mind, but a genuine Frenchman in sympathy and spirit, published a work on the history of modern chemical theories, which opens with this amusing and characteristic French panegyric: "Chemistry is a French science. It was founded by Lavoisier, of immortal memory;"

and the author goes on to make good his claim that a large part of the great generalizations in the science have been made by Frenchmen. This unmeasured assumption, coming, as it did, on the eve of the Franco-Prussian War, was the occasion of no little bitterness, and was answered in very much the same spirit in which it was uttered. The old controversies were revived, and the old arguments repeated, proving, what was undoubtedly true, that Lavoisier did not add a new fact of prime importance to chemistry. But he did add one of the grandest generalizations. Lavoisier was not a chemical investigator in our modern sense, but he had, to a very high degree, that quality of mind which the Frenchmen call *clarté,* and the great good fortune—if you please so to style the opportunities of genius—to advance his theory of combustion at the time the discoveries of Priestley, Cavendish, Black, and Scheele, had prepared the world to receive it. His contemporary, Scheele, a poor apothecary in an out-of-the-way village of Sweden, had done more than any one else to supply the facts which made the theory credible; but, not only did he not see clearly the bearing of his facts, but he had not the vantage-ground which would have enabled him to impress his ideas on his age; and, although, with his extremely restricted means, he added more knowledge to the stock of chemical science in a single year than did Lavoisier in his lifetime, yet it was Lavoisier, and not Scheele, who made the great generalization which revolutionized chemistry.

It is unnecessary to add that this Franco-German controversy was as irrational as it was useless. Both Lavoisier and Scheele filled well the place to which they were called, and did faithfully the work which Providence assigned them in the development of chemical science, and it is sheer presumption in any man to say that one was more important or more honorable than the other.

It is true that chemistry, as a science of exact *quantitative* relations, begins with the introduction of the balance into the science, and that Lavoisier was one of the first to recognize the importance of this instrument for investigating chemical problems. But, from the beginning of the seventeenth century, chemistry as a science of *qualitative* relations was actively studied at all the great centres of learning in Europe, and was illustrated by some of the most learned men of the age. For over a century previous to the time of Lavoisier, who died in 1794, the doctrines of the science centred around a theory of combustion which is known in history as the phlogiston theory. This theory was first advanced in 1682, by Becher, a German chemist then living in England, and was worked out into a complete system some years later by Stahl. According to this theory, the principle of fire is everywhere diffused throughout Nature, but enters into the composition of different bodies to a very unequal extent. Combustible substances are bodies very rich in phlogiston, and burning consists in the escape of phlogiston into the atmosphere. I have already referred to this theory, and shown that it was in variance with the great principle of the law of gravitation, that quantity of matter is proportional to weight. Still, as I said before, this principle of Newton made its way into chemistry very slowly, and the theory of Stahl was in complete accordance with the philosophy of Aristotle, which, at the time, held an entire supremacy over the intellectual world. And was the theory wholly false? I believe not; and I am persuaded that every theory, which gains among thinking men such universal acceptance as did this theory of Stahl, has its element of truth. The men of the seventeenth century were not less acute thinkers than ourselves, and we must be careful not to judge of their ideas from our stand-point. The authors of the theory never attached to phlogiston the idea of weight which we necessarily associate with all matter. It was to them a principle, an undefined essence, and not matter in the sense we understand it. Vague and indefinite idea, no doubt, like many of the metaphysical ideas of the time, but not absurd. And that it was not absurd a single consideration will show. Translate the word *phlogiston energy*, and in Stahl's work on chemistry and physics, of 1731, put *energy* where he wrote *phlogiston*, and you will find there the germs of our great modern doctrine of conservation of energy—one of the noblest products of human thought. It was not a mere fanciful speculation which ruled the scientific thought of Europe for a century and a half. It was a really grand generalization; but the generalization was given to the world clothed in such a material garb that it has required two centuries to unwrap the truth. Still, the sparkle of the gem was there, and men followed it until it led them into a clearer day. It is a great error to suppose that the theory of Lavoisier superseded that of Stahl. It merely added to it. Stahl clearly saw that the chief characteristic of burning was the development of energy, and, although he called energy phlogiston, and did not comprehend its real essence, he recognized that it was a fundamental principle of Nature. He did not understand the chemical change which takes place in the process, and this Lavoisier discovered. But, both Lavoisier and his followers, to a great extent, ignored the more important phenomenon in magnifying the less, and it is only within a few years that the true relations of the two have been understood. All honor to these great pioneers of science, and let their experience teach us that, in science, as in religion, we see as through a glass darkly, and that we must not attach too much importance to the forms of thought, which, like all things human, are subject to limitations, and liable to change.

LECTURE XI.

QUANTIVALENCE AND METATHESIS—ALKALIES AND ACIDS.

Before studying metathesis, the third, as you will remember, of the three classes into which we divided chemical reactions, I must ask your attention, at the beginning of my lecture this evening, to a most important general principle, to which a study of the results of analysis and synthesis has led, and which will greatly help to elucidate the metathetical processes we have yet to investigate.

HCl	H_2O	H_3N	H_4C		
$NaCl$	$HgCl_2$	$SbCl_3$	CCl_4	PCl_5	CrF_6
H_2O	HgO	$NOCl$	CO_2	$POCl_3$	CrO_3
			$COCl_2$		CrO_2Cl_2
			COH_2		

The diagram on the curtain before us illustrates the truth we have to present. The story, indeed, is here told in our chemical hieroglyphics, but let us try to decipher them. In attacking our work, let us not fail to remember that these symbols really exhibit the constitution of the molecules of the definite substances they represent. The symbol H_2O, for example, shows that a molecule of water consists of two atoms of hydrogen and one of oxygen. Remember that this symbol is not the expression of a mere hypothesis, but represents the results of actual experiment. In a former lecture we have dwelt at length on the evidence on which it is based. We cannot continually retrace our steps; but be sure that you recall this evidence, so that we may plant the ladder, on which we shall attempt to climb higher, on firm ground. Now, what is true of the symbol of water, is true of all the symbols on this diagram. There is not one of them in regard to which there is a shade of doubt. Our atoms may be mere fancies, I admit, but, like the magnitudes we call waves of light, the magnitudes we have measured and called atoms must be magnitudes of something, however greatly our conceptions in regard to that something may change. Our whole atomic theory may pass, the words molecule and atom may be forgotten; but it will never cease to be true that the magnitude which we now call a molecule of water consists of two of the magnitudes which, in the year 1872, were called atoms of hydrogen, and of one of the magnitudes which were called, at the same period, atoms of oxygen.

Look, now, at the first line of symbols, and see in what a remarkable relation the atoms there represented stand to each other. In a molecule of hydrochloric-acid gas (HCl), one atom of chlorine is united to one atom of hydrogen. In the molecule of water (H_2O) one atom of oxygen is united to two of hydrogen. In the molecule of ammonia gas (NH_3) one atom of nitrogen is united to three atoms of hydrogen, and in the molecule of marsh gas (CH_4) the atom of carbon is united to four atoms of hydrogen. It would appear, then, that the atoms of chlorine, oxygen, nitrogen, and carbon, have different powers of combination, uniting respectively with one, two, three, and four atoms of hydrogen. In order to assure yourselves that this relation is not an illusion, depending on the collocation of selected symbols, but results from a definite quality of the several atoms, examine the symbols of the second line, and you will see that, in a similar way, the atoms of sodium (Na), mercury (Hg), antimony (Sb), carbon (C), and phosphorus (P), unite respectively with one, two, three, four, and five atoms of chlorine. Moreover, on comparing the two lines, notice that the atom of chlorine, which combines with one atom of hydrogen, combines also with one atom of sodium. Again notice that the atom of carbon, which combines with four atoms of hydrogen, combines also with four atoms of chlorine. Further, observe on the third line that the atom of mercury, which combines with two atoms of chlorine, combines with only one of oxygen; and that the atom of carbon, which combines with either four atoms of chlorine or four atoms of hydrogen, combines with two atoms of oxygen; and compare with these facts those first noticed, that the atom of oxygen combines with two atoms of hydrogen, and the atom of chlorine with but one.

Relations so far reaching and so intricate as these cannot be accidental; and when you are told that the examples here given have been selected, on account of their simplicity, from a countless number of instances in which similar relations have been observed, you will not be satisfied until you find some explanation of the cause of these facts.

The explanation which our modern chemistry gives

is this: It is assumed that each of the elementary atoms has a certain definite number of bonds, and that by these alone it can be united to other atoms. If you wish to clothe this abstract idea in a material conception, picture these bonds as so many hooks, or, what is probably nearer the truth, regard them as poles like those of a magnet. If we have grasped this idea, let us turn back to our diagram and we shall find that the relations we had but dimly seen have become clear and intelligible. The hydrogen, sodium and chlorine atoms have only one bond or pole, and hence, in combining with each other, they can only unite in pairs. The oxygen-atom has two bonds or poles, and can combine, therefore, with two hydrogen-atoms, one at each pole. The mercury-atom has also two bonds, and takes, in a similar manner, two atoms of chlorine; but it can only combine with a single atom of oxygen, for the two poles of one just satisfy the two poles of the other. Again, the atom of carbon has four bonds, which may be satisfied by either four atoms of hydrogen, or four atoms of chlorine, or two atoms of oxygen, or one atom of oxygen and two of chlorine, or, lastly, one atom of oxygen and two of hydrogen. Further, the atom of phosphorus has five bonds, and holds five atoms of chlorine, or three atoms of chlorine and one of oxygen. Finally, the chromium atom binds six atoms of fluorine, or three of oxygen, or two of oxygen and two of chlorine. This quality of the atoms, which we endeavor to represent to our minds by the conception of hooks, bonds, or poles, we call, in our modern chemistry, quantivalence, and we use the Latin terms univalent, bivalent, trivalent, quadrivalent, quinquivalent, sexivalent, etc., to designate the atoms which have one, two, three, four, five, six, etc., hooks, bonds, or poles, respectively.

```
                        |          |
H –      – O –      – N –      – C –
                                   |

                        |          |
Cl –     – S –      – P –      – Si –
                                   |

                        |          |
F –      – Ca –     – Sb –     – Sn –
                                   |

                        |          |
K –      – Mg –     – As –     – Ti –
                                   |

                        |          |
Na –     – Hg –     – B –      – Pt –
                                   |

                        |          |
Ag –     – Zn –     – Au –     – Zr –
                                   |
```

In the above diagram we have classified a few only of the more important elementary atoms according to their quantivalence, and the diagram also shows how, by a slight addition to our symbolical notation, we can indicate the number of bonds in each case. In writing symbols of molecules, a dash between two letters indicates the union of two bonds, and one bond or pole on each atom is then said to be closed. Two dashes indicate that two bonds on each atom are closed—and so with a larger number. The next diagram is in part a repetition of that on page 232, with the exception that the bonds are indicated.

```
                                 H                H
                                 |                |
H – H      H – O – H      H – N – H      H – C – H
                                                  |
                                                  H

                                 Cl               Cl
                                 |                |
H – Cl     Cl – Hg – Cl   Cl – Sb – Cl   Cl – C – Cl
                                                  |
                                                  Cl

           Hg = O         Cl – N = O     O = C = O
```

You notice that this idea of quantivalence suggests, or, rather, as I should say, implies the idea that the molecules have a definite structure. Thus in the molecule CH_4 we conceive that the carbon-atom is united at four distinct points with the four hydrogen-atoms. There is not an indiscriminate grouping of the five atoms, but a definite arrangement with the carbon-atom at the centre of the system. So, also, in CCl_4, which has the same structure as CH_4, determined, as before, by the quadrivalence of the nucleus. Passing next to CO_2 we find an equally definite structure, the four bonds of the same nucleus being satisfied by two bivalent atoms of oxygen; and intermediate in structure, between the two molecules last mentioned, we have the molecule of phosgene gas, $COCl_2$, and the molecule of formic aldehyde, COH_2.

The symbols of these molecules indicate an obvious limitation to this idea of structure, which must not be overlooked, and which cannot too early be called to your notice. All that we, as yet, feel justified in inferring from the phenomena we have described, are simply the facts that in the molecule CCl_4, for example, the four chlorine-atoms are united to the carbon-nucleus by four different bonds, and that in the molecule CO_2 the two oxygen-atoms are united to the same nucleus, each by two bonds. Further than this we assert nothing. It may hereafter appear that the different bonds of the carbon-atom have different values; or, perhaps, have a fixed position, and that there are distinctions of right and left, top and bottom, or the like; but, until we are acquainted with phenomena which require assumptions of this sort, we may group our symbols around the nucleus of the molecule as we find most convenient, provided only we satisfy the condition of quantivalence. Thus it is unimportant whether we write

$$\mathrm{Cl-Hg-Cl}, \quad \text{or} \quad \mathrm{Hg}\begin{smallmatrix}\diagup \mathrm{Cl} \\ \diagdown \mathrm{Cl}\end{smallmatrix}; \quad \begin{array}{c} \mathrm{Cl-C-Cl}, \\ \Vert \\ \mathrm{O} \end{array} \quad \text{or} \quad \begin{array}{r} \mathrm{Cl} \\ | \\ \mathrm{O=C} \\ | \\ \mathrm{Cl}. \end{array}$$

The quantivalence of the atoms, moreover, is by no means an invariable quality; but this circumstance does not in the least obscure the general principle we have been discussing: because, in the first place, any change in the quantivalence of an atom is accompanied with a change in all its chemical relations; and, in the second place, the change is circumscribed by definite limits, which are easily defined. This point will be best illustrated by a few examples.

When in a previous lecture, as an example of a synthetical process, we united ammonia gas with hydrochloric acid, there was a change in the quantivalence of the nitrogen-atom, from three to five, as will be seen on comparing the symbol of the first factor with the sole product of the reaction:

$$\begin{array}{ccc} & & H \\ & & | \\ H & - & N \\ & & | \\ & & H \end{array} \qquad\qquad \begin{array}{cccc} & & H & \\ & & | & \\ H\diagdown & & & \\ & & N & - Cl \\ H\diagup & & & \\ & & | & \\ & & H & \end{array}$$

Ammonia Gas. Ammonic Chloride.

Now, from ammonia gas can be derived a large class of compounds, in all of which nitrogen is trivalent; and, in like manner, from ammonic chloride can be derived another class of compounds, in which nitrogen is quinquivalent; but, although they all contain the same atom as a nucleus, the two classes differ from each other as widely as if they were composed of different elements. A similar fact is true of phosphorus, which forms two well-marked chlorides:

$$\begin{array}{ccc} & & Cl \\ & & | \\ Cl & - & P \\ & & | \\ & & Cl \end{array} \qquad\qquad \begin{array}{cccc} & & Cl & \\ & & | & \\ Cl\diagdown & & & \\ & & P & - Cl \\ Cl\diagup & & & \\ & & | & \\ & & Cl & \end{array}$$

Phosphorous Chloride. Phosphoric Chloride.

One of the most striking instances of the variation of quantivalence is to be found in the atom of manganese. This elementary substance forms no less than four compounds with fluorine, whose molecules have probably the constitution represented by the symbols given below:

$$\mathrm{F - Mn - F} \qquad \begin{array}{c} \mathrm{F} \\ | \\ \mathrm{F - Mn - F} \\ | \\ \mathrm{F} \end{array} \qquad \begin{array}{ccccccc} & & \mathrm{F} & & \mathrm{F} & & \\ & & | & & | & & \\ \mathrm{F} & - & \mathrm{Mn} & - & \mathrm{Mn} & - & \mathrm{F} \\ & & | & & | & & \\ & & \mathrm{F} & & \mathrm{F} & & \end{array}$$

$$\begin{array}{c} \mathrm{F} \quad \mathrm{F} \\ \diagdown \ \diagup \\ \mathrm{F - Mn - F} \\ \diagup \ \diagdown \\ \mathrm{F} \quad \mathrm{F} \end{array}$$

In the first, the manganese-atom is bivalent; in the second and third it is quadrivalent; and in the last, sexivalent. The third molecule, it will be noticed, contains two quadrivalent atoms of manganese, united by a single bond, and the two together form a complex nucleus, which is sexivalent. Here, as in the previous examples, it is true that there is a distinct class of compounds corresponding to each of the four conditions of the nucleus, and that the difference between the chemical relations of the bivalent and those of the sexivalent atom of manganese is almost as great as that between the atom of zinc and the atom of sulphur.

The compounds of iron furnish a more familiar example of the effect produced by a variation of quantivalence, than either of those which have been adduced. There are two classes of these compounds, which are distinguished in chemistry as the ferrous and the ferric compounds. The first class consists of molecules, of which the nucleus is a bivalent atom of iron, while the molecules of the second class are grouped around a nucleus, consisting of two quadrivalent atoms united as explained above. Thus the symbols of ferrous and ferric chloride are:

$$FeCl_2 \text{ or } Cl-Fe-Cl, \text{ and } Fe_2Cl_6 \text{ or } \begin{array}{c} \quad\; Cl \quad\;\; Cl \\ \quad\;\; | \qquad\; | \\ Cl-Fe-Fe-Cl. \\ \quad\;\; | \qquad\; | \\ \quad\; Cl \quad\;\; Cl. \end{array}$$

Now, I have before me four glasses, which contain solutions in water of

$FeCl_2$,	Fe_2Cl_6,	$CuCl_2$		$NiCl_2$,
Ferrous Chloride,	Ferric Chloride,	Cupric Chloride,	and	Nickel Chloride;

and I will add to each glass a portion of a solution of a yellow salt, which is well known in commerce, under the name of yellow prussiate of potash, and in chemistry as potassic ferrocyanide. Notice, in the first place, what a different effect the reagent produces on the last two solutions. From the solution of cupric chloride, we obtain a red precipitate, and, from the solution of nickel chloride, a white precipitate. Next, we will add the same reagent to the solutions of the two compounds of iron, and, as you see, the difference of effect produced is even greater than before. Moreover, if, going behind the outward manifestations, you study the constitution of the products formed, you will find that the variations of color correspond to more fundamental differences in the case of the two conditions of iron than in that of the two separate elements, copper and nickel. The result, then, at which we arrive, is this, that, although a fixed quantivalence is not an invariable of quality of every atom, it is at least an invariable quality of each condition of every given atom, and that, in every marked class of compounds of any elementary substance, the atoms of that element always have the same quantivalence.

Lastly, as to the limits to which this variation of quantivalence may extend. There are several of the chemical elements, and these among the most important and most widely distributed, whose quantivalence appears to be invariable. This is especially true of hydrogen, it is likewise true of the alkaline metals, lithium, sodium, potassium, cæsium, and rubidium, and it is also true of silver, all elements whose atoms are univalent. It is further true of the trivalent element boron. Again, oxygen is always bivalent, and so are also the metallic radicals of the alkaline earths, calcium, barium, strontium, and magnesium, and so are, moreover, the well-known metallic elements, lead, zinc, and cadmium. Lastly, aluminum, titanium, silicon, and carbon, are always quadrivalent, although, in the single instance of the molecule, CO, the carbon-atom appears to be bivalent.

But, in addition to the fact that the variations in quantivalence are confined to a limited number of the elementary atoms, these variations appear to follow a remarkable law, which is thought to point to an explanation of their cause. As is shown in this diagram, the successive degrees of quantivalence in gold and phosphorus follow the order of the odd number:

$AuCl$	$AuCl_3$	
	PCl_3	PCl_5

while those of manganese follow the order of the even numbers:

MnF_2 MnF_4 MnF_6

Now, what is true of these atoms is, in general, true of the atoms of all those elements which have several degrees of quantivalence: at each successive step the quantivalence increases by two bonds, and never by a single bond. The explanation of the fact is thought to be that the bonds of any atom, when not in use to hold other atoms, are satisfied by each other, and that, so far as these unused bonds are concerned, the atom is in the condition of a horseshoe magnet, with its north pole directed toward and neutralized by its south pole. Thus it is assumed that, in both of the two compounds of carbon and oxygen, the carbon atom is quadrivalent, the only difference being that, while in CO_2 all four bonds are employed to hold the two atoms of oxygen, in CO only two are so used, the other two neutralizing each other thus:

O=C=O ⊂C=O.

Of course, then, if the unused bonds are in all cases neutralized in this way, it must be that the quantivalence of an atom will fall off from the highest degree of which it it susceptible, by two bonds at each step; so that, if the highest degree is odd, all must be odd, and, if the highest is even, all must be even, as in the illustrations given above. Atoms with odd degrees of quantivalence have been called perissads, and those with even degrees have been called artiads, and the classification appears to be a fundamental one; but there are important exceptions to the general principle, which have never yet been reconciled with the theory.

The doctrine of quantivalence, which we have endeavored to illustrate in this lecture, is one of the distinctive features in which the new chemistry differs from the old, and the recognition of the fact that a definite quantivalence is an inherent quality of each elementary atom was one of the chief causes of the revolution in the science which has recently taken place. In the old chemistry, the question of how the elementary substances were united in a compound was hardly raised, much less answered; but now the manner in which the atoms are grouped together in the molecule has become an all-important question. Every molecule is a unit in which all the atoms are joined together by their several bonds, and it becomes an object of investigation to determine the exact manner in which the molecular structure is built up. Moreover, it appears that the qualities and chemical relations of a compound are determined fully as much by the structure of its molecules as by the nature of the atoms of which the molecules consist. For example, it was formerly supposed that the qualities of an alkali or an acid were simply the characteristics of the compounds of certain elements with oxygen, but it now appears that they are the result of a definite molecular structure, and are only slightly modified by the characteristics of the individual atoms which may chance to be the nucleus of the molecule.

We are thus fairly brought face to face with the question of molecular structure that is to occupy our attention during the remainder of this course of lectures. In regard to this question, there are a few preliminary points which need barely be mentioned, as they can easily be apprehended, and require, therefore, no extended illustration. It is evident that with univalent atoms solely we can only form molecules consisting of two atoms, like Na-Cl, or H-Br. When we introduce bivalent atoms the structure becomes more complex—as in H-O-H or K-O-Cl. With several bivalent atoms we can form molecules in which the atoms seem to be strung together in a chain, sometimes of great extent, as—

H - O - Ca - O - H, or H - O - Pb - O - Pb - O - Pb - O - H.
Calcic Hydrate. Triplumbic Hydrate.

And, with atoms of higher quantivalence, we obtain groups of very great complexity, of which the multivalent atom[1] is the nucleus, and serves to bind together

[1] The atom with a high degree of quantivalence.

the parts of the molecule. The molecule of calcic sulphate, for example, is supposed to have the complex con-

```
   /O\  //O
Ca      S
   \O/  \\O
```

Calcic Sulphate.

stitution which our symbol indicates, and it will be seen that it is the sexivalent atom of sulphur, which is the nucleus of the group, and holds the atoms together. So, also, in the still more complex molecule of alum, the double atom of aluminum is the nucleus of the group,

```
                 O   O
                  \\ //
                   S
                  / \
      O          O   O          O
      ||         |   |          ||
K - O - S - O - Al - Al - O - S - O - K
      ||         |   |          ||
      O          O   O          O
                  \ /
                   S
                  // \\
                 O   O
```

Potassic-Aluminic Sulphate (Alum).

and unites the several parts, while the four sexivalent atoms of sulphur are the centres of subordinate groups connected with this nucleus. Notice that all the atoms are united by their respective bonds, and that to each set is assigned a definite quantivalence, and you can hardly fail to appreciate the important fundamental principles of our modern chemistry, which I have been endeavoring to illustrate. They may be summed up in the following terms:

The integrity of every complex molecule depends on the multivalence of one or more of its atoms, and no such molecule can exist unless its parts are bound together by these atomic clamps.

Such symbols as those just given, by which we attempt to indicate the relations of the parts of a molecule, are called graphic or sometimes rational symbols, and are to be distinguished from those we have hitherto used, which, as they represent simply the results of experiment, are known as empirical symbols. Of course, these graphic symbols are the expressions of our theoretical conceptions, and must survive or perish with the theory that gave them birth. But, absurd as these conceptions certainly would be if we supposed them realized in the concrete forms which our diagrams embody, yet, when regarded as aids to the attainment of general truths, which in their essence are still incomprehensible, these crude and mechanical ideals have the greatest value, and become very important aids to the study of chemical science.

The molecular structure of bodies is inferred chiefly from the reactions of which they are susceptible, or by which they are formed, and I now propose to ask you to study with me a number of chemical processes which I have selected with a view of illustrating the structure of a few of the more important classes of chemical compounds. The processes best adapted for our purpose, and therefore selected, are chiefly examples of metathesis, and incidentally we shall become acquainted with this third class of chemical reactions.

Metathesis consists in the interchange of atoms or groups of atoms between two molecules, and implies that the structure of these molecules is not otherwise altered. Such an interchange, of course, involves the breaking up of two sets of molecules, and the production of two new sets, and might be regarded as a concurrence of analysis and synthesis; but the cases are so very common of chemical processes in which one atom or a group of atoms is simply substituted for another, without otherwise altering the structure of the molecules concerned, that it is convenient to study these reactions by themselves. The first example which I shall bring to your notice is the reaction of metallic sodium on water.

The effect of pure sodium on water is so violent that we find it convenient to moderate the action by amalgamating the metal with mercury, which, without in the least degree altering the relations of the sodium to the water, reduces the rapidity of the chemical process. We will, now, pass under this glass bell, which is filled with water, and standing on the shelf of the pneumatic trough, a bit of this sodium amalgam. You notice a rapid evolution of gas, which soon nearly fills the bell. Let us examine this gas. On bringing the open mouth of the bell near a candle-flame, the gas takes fire and burns with the familiar appearance of hydrogen, and this is sufficient to assure you that the product with which we are here dealing is hydrogen gas. But what is the other product of the reaction? To discover this, we will next place another lump, this time of the pure metal, on an open pan of water. The metal is lighter than water, and floats on the surface, but, to prevent it from swimming around, we have placed on the liquid a disk of filtering-paper on which the lump rests. The action is now very violent—hydrogen gas is evolved as before, a high temperature is developed, and the metal melts. Were the melted globule free to swim on the surface of the water, the cooling effect of the liquid would prevent the temperature from rising to the point of ignition. As it is, however, the globule being entangled by the filter-paper, the heat soon accumulates to a sufficient degree to inflame the escaping hydrogen, and this furnishes us with the evidence that the gas is really escaping. But the color of the flame is not familiar.

Why is it so yellow? Simply because the flame contains a small amount of the vapor of sodium, and the merest trace of that vapor in any flame is sufficient to color it intensely yellow. Any volatile compound of sodium introduced into a non-luminous gas-flame produces the same effect. But where is that other product we are seeking? Evidently we must look for it in the water, on which the sodium has been acting. Have the qualities of the liquid changed? This question can be answered by a simple test. Here we have some strips of paper, which are colored with certain well-known vegetable dyes. The yellow strips are colored with turmeric, and the red with litmus. On dipping these strips in a jar of pure water, notice that the color is not in the least degree modified; but mark that, when the yellow strip is drawn through the water on which the sodium has been acting, the color becomes at once bright red; while, on the other hand, the strip colored red by litmus becomes blue. Evidently it is some product of the reaction dissolved in the water which produces these changes, and this conclusion will be confirmed on tasting the water, which has acquired a sharp, biting taste, and attacks the skin, producing, when rubbed between the fingers, a peculiar unctuous feeling, effects which every one will recognize as those of a caustic alkali. If, now, we evaporate the water, we shall obtain a small quantity of an amorphous, white solid, similar to that which is contained in this bottle, and which is only a purer form of the caustic soda of commerce used in such great quantities for making soap.

As we are able to discover no other results of this process except the two substances you have seen, you may conclude that the only products of the reaction of sodium on water are hydrogen gas and caustic soda.

Next, as to the nature of the process, and how we can express it by our symbols. We know all about the molecular constitution of the factors of the reaction. The symbol of a molecule of sodium is Na-Na, and that of water H-O-H. These molecules have the simplest types of structure. We also know that the molecule of hydrogen gas has the symbol H-H, but how about the molecule of caustic soda (sodic hydrate, as we call it)? Chemical analysis shows that this substance consists simply of sodium, oxygen, and hydrogen, in proportions, by weight, corresponding exactly with those proportions which have been assumed to be the relative weights of the atoms of these three elements. Analysis, therefore, proves that the molecule of caustic soda contains an equal number of atoms of all three of its elementary constituents, but it does not enable us to decide whether its symbol is NaOH or $Na_2O_2H_2$, or any other simple multiple of these letters. Here, however, the principles of quantivalence come to our aid. We know that both H and Na are univalent atoms, and that the molecule of oxygen can only hold two such atoms. Hence the symbol must be Na-O-H, and can be nothing else. Were caustic soda a volatile solid, so that we could determine the specific gravity of its vapor, we could reach a knowledge of its molecular constitution in the manner previously described, which is much more direct and satisfactory; but, as it cannot be volatilized within any manageable limits of temperature, we are obliged to resort to methods whose results are undoubtedly less conclusive, and depend, to a greater or less degree, on theoretical considerations.

Writing out, now, the symbols of the factors and products of our reaction,

Na-Na H-O-H Na-O-H H-H,

we notice that, as there are two atoms of Na in the molecule of the metal, we must have formed two molecules of Na-O-H, and, as there will then be four atoms of hydrogen among the products, there must be two molecules of water used in the factors, and our reaction, thus amended, becomes

$$\text{Na-Na} + 2\text{H-O-H} = 2\text{Na-O-H} + \text{H-H}.$$

If, next, we represent the reaction by graphic symbols, the nature of the change will be made still more evident:

$$\begin{matrix}\text{H}-\text{O}-\text{H}\\ \text{H}-\text{O}-\text{H}\end{matrix} + \begin{matrix}\text{Na}\\ |\\ \text{Na}\end{matrix} = \begin{matrix}\text{Na}-\text{O}-\text{H}\\ \text{Na}-\text{O}-\text{H}\end{matrix} + \begin{matrix}\text{H}\\ |\\ \text{H}.\end{matrix}$$

It will be now seen that the two atoms of sodium have changed place each with an atom of hydrogen in the molecule of water, and that the displaced atoms of hydrogen have taken the place of the atoms of sodium. In a word, the new molecules have precisely the same structure as the old, and only differ from them in the substitution of Na for H, or the reverse. This reaction is, therefore, a simple example of metathesis.

Caustic soda (or sodic hydrate), which was one of the products of the reaction we have been studying, belongs to a class of substances which have long been distinguished for their very marked and useful qualities, and are called alkalies. The most striking and familiar of these qualities have already been noticed, and, among others, the effects which the alkalies produce on the colored papers dyed with turmeric or litmus. Now, there is another class of compounds whose qualities, while equally marked, bear a most striking antithesis to those of the alkalies. These compounds are called acids, and the word recalls a peculiar taste and a corrosive action, with which every one is more or less familiar. Here we have one of these substances, the muriatic acid of commerce, which, as I have already told you, is a solution of hydrochloric-acid gas (HCl) in water. Notice that, when I dip in this acid solution the dyed papers which have been altered by the alkali, their former color is at once restored. The acid thus undoes the effect of the alkali, and, what is more, if I add the acid slowly to the alkaline solution, and, after each addition, test the solution with my papers, I shall find that the alkaline reaction, as we call it, becomes feebler and feebler until at last it wholly disappears. So, on the other hand, if we add the alkaline to the acid solution, the test-papers will show that the acid qualities disappear in a similar manner, and we can easily bring the solution to such a condition that it has no more effect on the vegetable dyes than so much pure water. This chemical process is usually described by saying that the acid and alkali neutralize each other, and notice that in the case before us the test-papers show that the neutral point has been reached. On tasting the solution, we cannot discover the least traces of either an acid or an alkaline taste, but in their place we recognize the flavor of common salt, and if we evaporate the solution we shall obtain a small quantity of this most familiar condiment.

With all the substances concerned in the reaction we have just studied, we are perfectly familiar. Let us see, then, if we cannot express the reaction by means of our chemical symbols:

$$\underset{\text{Sodic Hydrate.}}{\text{Na-O-H}} + \underset{\text{Hydrochloric Acid.}}{\text{HCl}} = \underset{\text{Water.}}{\text{H-O-H}} + \underset{\text{Sodic Chloride.}}{\text{Na-Cl.}}$$

The reaction evidently consists in the simple substitution of Na for H in the molecule of HCl, and the reproduction of a molecule of water, which, mixing with the great mass of water present, would naturally be lost sight of in the experiment.

It appears, then, that, in the present case at least, the neutralizing of an acid by an alkali is a simple metathetical reaction, in which the metallic atom of the alkaline-molecule changes place with the hydrogen-atom of the acid-molecule. Now, the chief interest of this experiment arises from the fact that it is a single example of a general truth, and the principle is one of such importance that it requires further illustration.

On the second pan of water I therefore throw a lump of another metallic element, closely allied to sodium, called potassium. The action is even more violent than before, and mark that the escaping hydrogen inflames while the metallic globule is swimming rapidly about on the surface of the water. Notice, also, the beautiful color which the potassium-vapor imparts to the flame, so different from that obtained with sodium. These colors are, in fact, very characteristic, and, when examined with the spectroscope, are condensed in certain luminous bands, whose positions on the scale of the instrument afford a never-failing indication of the presence of the metal in the flame. You see, moreover, that, as before, the water has acquired an alkaline reaction, and, if we evaporate the solution, we shall obtain a small quantity of a white solid called potash (or potassic hydrate), so similar to caustic soda that the two can scarcely be distinguished except by chemical tests. The process is so analogous, in every respect, to the last, that it is certainly unnecessary to repeat the evidence on which our knowledge of the reaction is based, but we will express it at once by our chemical symbols:

K-K	+	2H-O-H	=	H-H	+	2K-O-H.
Potassium.		Water.		Hydrogen Gas.		Potassic Hydrate.

The sole difference is that we have here atoms of potassium, K, instead of atoms of sodium, Na, which, however, like the last, take the place each of a hydrogen-atom in one of the molecules of water.

In the previous example we neutralized the alkali soda with hydrochloric acid. We have here another compound of the same class, called nitric acid, and let us see whether, in like manner, this acid will neutralize the alkali potash. Notice that, as we add the acid, the alkaline reaction becomes feebler and feebler, until at last it has entirely disappeared. The liquid has now no effect on either of these sensitive papers. On tasting it, we discover no pungency, and likewise no acidity, but we recognize a peculiar saline taste, which is not unfamiliar. Here is a bit of paper which has been dipped in a similar solution and dried. See how it sparkles when lighted, and every boy will tell us that we are dealing with the well-known salt we call nitre. And so it is; and, on evaporating the solution, we should obtain the familiar crystals of this substance.

Before we can explain this new reaction, we must know what is the symbol of a molecule of nitric acid, and also that of a molecule of nitre. Since neither of these substances can be volatilized without decomposition, we cannot weigh their vapor, and cannot therefore apply the method of finding the symbol we explained in a previous lecture. As in the case of the sodic hydrate, however, we are not wholly helpless, for analysis will tell us a great deal, and, once for all, let us consider just how much information an accurate analysis will give us in regard to the symbol, and how far it leaves us in the dark.

Here, then, we have the analysis of nitric acid, and in regard to the accuracy of these numbers there cannot be a doubt:

Hydrogen....................	2.57	1	H
Nitrogen....................	35.89	14	N
Oxygen......................	61.54	48	O_3
	100		

Nitric acid consists of the three elementary substances —hydrogen, nitrogen, and oxygen—in the exact proportions here indicated, just so many per cent. of each. Now, these per cents. are to each other precisely as the numbers 1 : 14 : 48 ; or, as the weight of one atom of hydrogen is to the weight of one atom of nitrogen is to the weight of three atoms of oxygen ; or, in symbols, as $H : N : O_3$. But, as every one knows, we may multiply all the terms of a proportion by any number we please without in the least altering the value of the ratios—thus, $1 : 14 : 48 = H : N : O_3 = H_2 : N_2 : O_6 = H_3 : N_3 : O_9$; or, in general, as $H_n : N_n : O_{3n}$. Hence, then, if nitric acid consists of hydrogen, nitrogen, and oxygen, in the proportions which our analysis indicates, its molecule must be represented either by HNO_3, or by some simple multiple of these symbols. Knowing, then, as we do, the relative weights of the atoms, simple analysis will tell us in every case the relative number of atoms present in the molecule, but it cannot fix the absolute number.

You see, therefore, that analysis alone gives us always a close approximation to the symbol, and limits the question within very restricted bounds. The simplest formula in any case is that which represents the molecule as consisting of the smallest number of whole atoms which will satisfy the conditions, and the only question can be as between this symbol and its multiples. In the case of all volatile compounds, a very rough determination of their vapor density is sufficient to decide the question. Thus, in the case of nitric acid, if the symbol is HNO_3, the molecular weight is 63, and the vapor density would be 31.5. Were the symbol $H_2N_2O_6$, the density would be 63; were it $H_3N_3O_9$, the density would be 94.5; and, although there are causes which make many of our determinations of vapor densities untrustworthy within several per cent., they are abundantly accurate enough to show which of such widely-differing values must be the true one.

Hence, although theoretically the molecular weight, as determined by the vapor density, is our starting-point in the investigation of the symbol of a compound, practically it is only used to control the results of analysis. So also, when, in the case of non-volatile compounds, we must resort to other modes of fixing the molecular weight, an accurate analysis having once been made, the question lies only between a few widely-differing numbers, and considerations are sufficient to decide between these which would not be regarded as satisfactory were greater accuracy required.

Of course, as must be expected, there are substances in regard to which no definite conclusions can be reached, and where conflicting evidence renders differences of opinion possible. This is true of many mineral species, and the symbols of such compounds are in doubt to the extent I have mentioned. In such cases, we usually adopt provisionally the simplest symbol, and wait for the advance of science to correct any error which may be made, and which, for the time at least, is unimportant.

This is, undoubtedly, a difficult subject—one of the most difficult in chemistry; but the difficulty can be mastered with a little thought, and it requires no detailed knowledge of the science to follow the reasoning thus far. It is different, however, with the purely chemical evidence on which we are frequently obliged to rely for deciding between the few formulas which, in a given case, analysis shows may be possible. This evidence will have no force, except with those who have already a competent knowledge of the facts. Thus much, however, can be understood. The facts of chemistry, like those of any other science, are parts of a general plan more or less fully apprehended by the student, and the evidence of which I am speaking may be summed up in the statement that the given symbol is accepted because it is consistent with this plan. Of course, such reasoning is not absolutely conclusive, and there is room for doubt, but so there is in every department of science. A part of the way we walk in the clear light of knowledge; the rest of the way we grope; but it is only thus that we can penetrate the darkness of the unknown, and we rely on that intelligence in man which finds its response in the intelligence of Nature, to direct our steps.

Having now explained as fully as our time will permit the general nature of the evidence on which we depend for establishing the empirical symbol of a compound, I shall not recur to it again, but shall regard it as sufficient to say that chemists are agreed that the symbol is thus or so. In the case of nitric acid, there is no question that the symbol of the molecule is HNO_3, and, in like manner, KNO_3 is the received symbol of nitre. How, now, shall we write the reaction we last studied? Simply thus:

$$\underset{\text{Potassic Hydrate.}}{K\text{-}O\text{-}H} + \underset{\text{Nitric Acid.}}{H\text{-}NO_3} = \underset{\text{Water.}}{H\text{-}O\text{-}H} + \underset{\text{Potassic Nitrate.}}{K\text{-}NO_3}.$$

The reaction, then, consists merely in an interchange between the hydrogen-atom of the acid and the metallic atom of the alkali. It is, then, precisely similar to the reaction between sodic hydrate and hydrochloric acid; and, if, as I said before, these are only examples of what is true in the case of all alkalies and all acids, we are certainly justified in deducing from our experiments the following principles: First, an alkali is a substance whose molecules have a definite structure, and differ from the molecules of water only in having a metallic atom in place of one of the hydrogen-atoms of the water-molecule; secondly, an acid is a substance whose molecules contain at least one atom of hydrogen, which is readily replaced by the metallic atom of the alkali when the two substances are brought together.

As the illustrations already given indicate, the characteristic qualities of an acid depend upon the circumstance that certain hydrogen-atoms in the molecules of these substances are readily replaced by metallic atoms. In my next lecture, I shall show that this susceptibility to replacement depends upon a definite molecular structure, but I must not leave this subject without insisting on the fact that this characteristic of acids is manifested in other ways besides the special mode we have been studying. A few experiments will illustrate this point:

In this flask there are some wrought-iron nails. We pour over them some muriatic acid, and warm the vessel. At once there is a brisk evolution of gas, which we are here collecting, in the usual way, over water; and notice that, when lighted, the gas burns with the familiar flame of hydrogen. Muriatic acid is an old friend. We know all about its constitution, and it is evident that the iron-atoms have replaced the hydrogen-atoms of the acid. If we evaporate the solution left in the flask, we shall obtain a green salt consisting of chlorine and iron. The reaction is thus represented:

$$\begin{array}{ccccccc} \begin{array}{c}\mathrm{H-Cl}\\ \\ \mathrm{H-Cl}\end{array} & + & \mathrm{Fe} & = & \mathrm{Fe}\begin{array}{l}\diagup\mathrm{Cl}\\ \\ \diagdown\mathrm{Cl}\end{array} & + & \begin{array}{c}\mathrm{H}\\ |\\ \mathrm{H}.\end{array} \\ \text{Hydrochloric Acid.} & & \text{Iron.} & & \text{Ferrous Chloride.} & & \text{Hydrogen Gas.} \end{array}$$

As the iron-atom is bivalent, it takes the place of two atoms of hydrogen, which, when thus displaced, form a molecule of hydrogen gas.

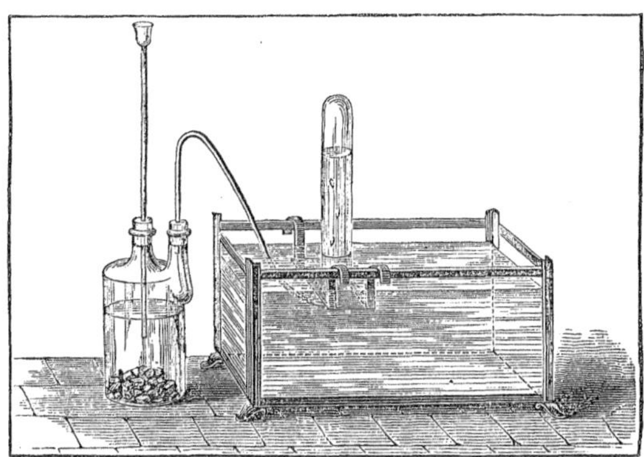

FIG. 31.—Preparation of Hydrogen Gas.

In the second flask are some zinc-clippings, and we will pour over them some dilute sulphuric acid, one of the best known of the class of compounds we are studying. Again, notice a brisk evolution of gas (Fig. 31), which also, as you see, burns like hydrogen. Indeed, this is the process by which hydrogen gas is usually made:

$$\begin{array}{ccccccc} \mathrm{H_2{=}SO_4} & + & \mathrm{Zn} & = & \mathrm{Zn{=}SO_4} & + & \mathrm{H{-}H.} \\ \text{Sulphuric Acid.} & & \text{Zinc.} & & \text{Zinc Sulphate.} & & \text{Hydrogen Gas.} \end{array}$$

In the reaction, which is here written, you notice that, as before, the metallic atom takes the place of two atoms of hydrogen; but sulphuric acid differs from both hydrochloric acid and nitric acid in that each of its molecules has two atoms of hydrogen, which can be thus replaced.

Examples like these might be multiplied indefinitely. We will conclude, however, with one more experiment, which illustrates the same susceptibility to substitution, but under slightly different conditions. This white powder is called zinc oxide, and is a compound of zinc with oxygen. Notice that it dissolves readily in a portion of the same dilute sulphuric acid used in the last experiment. Moreover, on evaporating the solution, we should obtain zinc sulphate ($ZnSO_4$), the same product as before. Why, then, is there no hydrogen gas evolved? Let our symbols tell us:

$$\underset{\text{Zinc Oxide.}}{ZnO} + \underset{\text{Sulphuric Acid.}}{H_2SO_4} = \underset{\text{Water.}}{H_2O} + \underset{\text{Zinc Sulphate.}}{ZnSO_4}.$$

You see that the metathesis yields water instead of hydrogen gas, and the question is answered.

LECTURE XII.

ELECTRO-CHEMICAL THEORY.

In our last lecture we saw that, whether an acid is brought in contact with an alkali, a metal, or a metallic oxide, one or more of the hydrogen-atoms in its molecules become replaced by metallic atoms from the molecules of the associated body, and this susceptibility to replacement was, as I stated, the distinguishing feature of that class of compounds we call acids. But I should leave you with a very imperfect notion of these important relations, if I did not proceed further to illustrate that the class of compounds we call alkalies, and which we have been accustomed to regard as the very opposite of acids, have exactly the same characteristics.

In this small glass flask there are some clippings of the metal aluminum, the metallic base of clay which has, within a few years, found many useful applications in the arts. On this metal I pour a solution of caustic potash. Notice that, on heating the flask, I obtain a brisk evolution of gas. On lighting the gas, it burns with a flame which leaves us no doubt that the gas is hydrogen. What, now, is the reaction? Somewhat more complex than those you have previously studied, because the atom of aluminum has a quantivalence of six. Moreover, in order to satisfy certain very striking analogies, we write the symbol of this atom Al_2, that is, we take 27.4 m.c. of aluminum for the assumed atom, and represent that by Al, although 54.8 m.c., which we write Al_2, is the smallest quantity of the element of which we have any knowledge, or which changes place with other atoms in the numerous metathetical reactions with which we are acquainted. Here

$$\begin{matrix} K-O-H \\ K-O-H \\ K-O-H \\ K-O-H \\ K-O-H \\ K-O-H \end{matrix} + Al_2 = K_6{}^{vi}O_6{}^{vi}Al_6 + 3H\text{-}H.$$

the Al_2 takes the place of six hydrogen-atoms, thus binding together what were before six distinct molecules of K-O-H into a single molecule of the resulting product. Evidently, then, the hydrogen-atom in the molecule of the alkali has the same facility of replacement as that in the molecule of the acid. Nor is this an isolated example, although, perhaps, the most striking we could adduce, and it illustrates a truth which was recognized long before the general adoption of the new philosophy of chemistry. Acids and alkalies belong to the same class of compounds, and caustic potash and nitric acid are simply the opposite extremes of a series of bodies in which all the intermediate gradations are fully represented. In our modern chemistry we call this class of chemical substances *hydrates*, and we distinguish the two extremes of the class as alkaline (or basic) and acid hydrates, respectively. The terms alkaline and basic are here used synonymously, although the first is generally restricted to the old caustic alkalies, including ammonia and the few compounds closely allied to them, which have been recently discovered.

Seeing, now, that the hydrogen-atom in the molecule of potassic hydrate has the same susceptibility of replacement whose cause we are seeking to discover, and knowing, as we do, the structure of this alkaline molecule, may it not be a similar structure which determines the like susceptibility in the molecules of all acids; for example, in those of nitric acid? What, now, is the position of the hydrogen-atom in the molecule which we have so often written, K-O-H? Why, simply this. It is one end of a chain of three atoms which has an atom of the metal potassium at the other end, and an atom of oxygen connecting the two. Now, we can write the symbol of nitric acid thus:

$$H-O-\left(N\begin{matrix}\nearrow O\\ \searrow O\end{matrix}\right),$$

and you will observe that we thus satisfy all the conditions of quantivalence, and have a structure similar to that of potassic hydrate. As before, we have an atom of oxygen uniting the hydrogen atom with the other end of the chain; but then this end of our molecular structure is formed, not by a single atom, but by a group of atoms (NO_2) which, nevertheless, can be replaced by metathesis just like a simple atom.

Allow me here, however, to make a short digression from the main line of my argument, in order to define an important term which we shall have frequent occasion to use during this lecture. By comparing the symbol K-O-H with H-O-(NO_2), it will be evident that the only essential difference between them is that the group NO_2 in the last takes the place of the atom K in the first. It must be, then, the influence of this part of the molecule which determines the difference between a strong alkali and a strong acid. Now, such an atom or such a group of atoms, which appears to determine the character of the molecule, is constantly called in chemistry a radical. Thus K is the radical of the molecule K-O-H, and NO_2 the radical of the molecule H-O-NO_2; but, while the potassium-atom is called a simple radical, the group NO_2 forms what is known as a compound radical. The influence of simple radicals in determining the qualities of their compounds has long been recognized. Indeed, the old chemistry laid altogether too much stress on this influence, regarding the qualities of a substance as derived in some unknown but remote manner from the qualities of its elements, and wholly ignoring the effect of molecular structure on these qualities, which we now know to be at least equally great. It was a very great step forward when the German chemist Liebig first recognized the truth that a group of atoms might give a distinctive character to a class of compounds just as effectively as an elementary atom. These groups he first named compound radicals, and assigned some of the names by which the more important of them are still known, and we now speak just as familiarly of the compounds of cyanogen (CN), of ammonium (NH_4), of methyl (CH_3), of ethyl (C_2H_5), etc., as we do of the compounds of chlorine, potassium, zinc, or iron. Moreover, each of the compound radicals, like a simple radical, has a definite quantivalence, but, while the quantivalence of the simple radical depends on wholly unknown conditions, that of the compound radical depends on the quantivalence of the elementary atoms of which it consists. Thus the radical NO_2 is univalent because one only of the five bonds of the nitrogen-atom remains unclosed, as the symbol indicates. Examine, also, the graphic symbols of the other compound radicals mentioned above:

$$\begin{array}{cccc} \begin{array}{c} H\;\;H \\ \diagdown\;\diagup \\ N- \\ \diagup\;\diagdown \\ H\;\;H \end{array} & -(C\equiv N) & \begin{array}{c} H \\ | \\ H-C- \\ | \\ H \end{array} & \begin{array}{c} H\;\;H \\ |\;\;\;| \\ H-C-C- \\ |\;\;\;| \\ H\;\;H \end{array} \\ \text{Ammonium.} & \text{Cyanogen.} & \text{Methyl.} & \text{Ethyl.} \end{array}$$

In each case, the number of bonds which are not closed determines the quantivalence.

Returning, now, to our comparison between K-O-H and H-O-NO_2, we should describe the relations of the molecules in a few words by saying that the acid and the alkali had molecules of the same general structure, but differed in that the radical of the alkali was the elementary atom potassium, while the radical of the acid was the atomic group NO_2.

As the result, then, of our discussion, we are led to the theory that acids and alkalies are compounds having the same general molecular structure, and that the susceptibility to replacement of the hydrogen-atom or atoms, which all these compounds contain, depends upon the molecular structure, while the differences between acids and alkalies, and, we might add, the differences between individual acids or individual alkalies, depends on the nature of the radical. Having been led thus far, the question next arises, Can we trace any connection between the acid and alkaline characters of the compounds, on the one side, and the nature of the radicals, which appear to determine these features, on the other side?

The simple radicals, as they appear in the elementary substances, may be divided into two great classes, the metals and the non-metals, the last class, by a singular perversion of language, being frequently called metalloids. Now, the most elementary knowledge of chemistry shows that, while radicals of opposite natures combine most eagerly together, two metals, or two closely-allied metalloids, show but little affinity for each other. These facts suggest at once an analogy between chemical affinity and the familiar manifestations of polar forces in electricity and magnetism; where it is also true that the like attracts the unlike. Moreover, it is found that, when, in the various processes of electrolysis, chemical compounds are decomposed by the electrical current, the different elementary substances appear at different poles of the electrical combination. Thus, hydrogen, potassium, and, in general, the metals, are evolved at what is called the negative pole, while oxygen, chlorine, bromine, and the allied metalloids, appear at the positive pole. It was, then, not unnatural to refer these effects of electrolysis to the electrical condition of the atoms, and to assume that the atoms had an opposite polarity to that of the poles, to which they were attracted, and hence the metals came to be called electro-positive and the metalloids electro-negative radicals; and these facts were thought very greatly to confirm the notion that chemical affinity is a manifestation of polar force closely allied to electrical attraction.

As expounded by the great Swedish chemist, Berzelius, this electro-chemical theory gave new life to that system of chemistry which, introduced into the science by Lavoisier and his contemporaries, has been only recently superseded. Corresponding to the duality of the electrical and magnetic poles, it was argued that there must be a duality in all chemical compounds, the elements uniting by twos to form binary compounds, the binaries again uniting by twos to form ternary compounds, and so on; and from this, its most characteristic feature, the old philosophy is now called the *dualistic system.* As the knowledge of chemical compounds has been enlarged, it has been found that, whatever may be the resemblances between electrical and chemical attraction, the analogy fails in the very point on which the dualistic system relied. But the chemists of the new school, in their reaction from dualism, have too much overlooked the electro-chemical facts, which are as true now as they ever were. The distinction between positive and negative radicals, based on their electrical relations, is evidently a most fundamental distinction, although, as Berzelius himself showed, the distinction is a relative and not an absolute one. It is possible to classify the radicals in one or more series in which any member is positive toward all that follow it, and negative toward all that precede it in the same series, and this principle is as true of the compound as it is of the simple radicals. Now, it is in this difference between positive and negative radicals that we shall find the origin of the distinctive features of the acid and the alkali.

Compare, again, the symbols of potassic hydrate and nitric acid as we have now learned to write them— K-O-H and $H-O-NO_2$—and seek, by the electro-chemical classification, to determine what are the electrical relations of the radicals K and NO_2, to which, as I have said, we must refer the distinctive features of these compounds. It will appear that K, the radical of the alkali, is the most highly electro-positive, and NO_2, the radical of the acid, one of the most highly electro-negative of all known radicals. Moreover, if you will extend your study, and compare in a similar manner the electrical relations of the other well-marked alkaline and acid hydrates, you will find that the radicals of the alkalies are all electro-positive, and the radicals of the acids all electro-negative, and, further, that the distinctive features of the alkali or the acid are the more marked in just the proportion that the position of the radical of the compound, in the electrical classification, is the more extreme. Lastly, those hydrates whose properties are indifferent, and which sometimes act as acids and sometimes as alkalies, will be found to contain radicals occupying an intermediate position in the same classification.

In following, then, the path which theoretical considerations have opened, we have met with a most remarkable class of facts. Alkalies contain radicals which, in the process of electrolysis, are attracted toward the negative pole of the battery, while acids contain radicals which, under the same conditions, are drawn toward the positive pole, and, in the proportion as the energy thus mutually exerted between radical and pole is the more marked, the acid or alkaline features of the hydrates of the radical are the more pronounced. Here are the facts, which no one will question; and what, now, is the explanation of them? We can give only a theoretical explanation based on the analogies of polar forces, a mode of manifestation of energy of which the chemical force appears to partake, as the very phenomena of electrolysis indicate.

If we carefully study what we have called the distinctive features of acids and alkalies, they will be found to depend on this, that the hydrogen-atoms of acids are readily replaced only by positive, and the hydrogen-atoms of alkalies only by negative radicals. In other words, with every hydrate the power of easily replacing its hydrogen-atom is only enjoyed by those radicals which are opposite in their electrical relations to the radical which the hydrate already contains. This will be found to be the one characteristic to which all that is peculiar to either acid or alkali can be referred, and if we can explain this we have explained all.

The explanation we would offer is as follows: The oxygen-atom with its two bonds, -O-, is in a condition similar to that of a bar of soft iron, susceptible of magnetism. When we unite the atom by one of these bonds with a positive radical, we produce an effect similar to that obtained by placing in contact with one end of such an iron bar a powerful magnetic pole. Under these conditions, as is well known, the two ends of the bar become strongly polar, the farther extremity acquiring a polarity of the same kind as that of the active pole; and so, in the case of our oxygen-atom, a positive radical united at one bond seems to polarize the atomic mass, and make a positive pole at its other end.

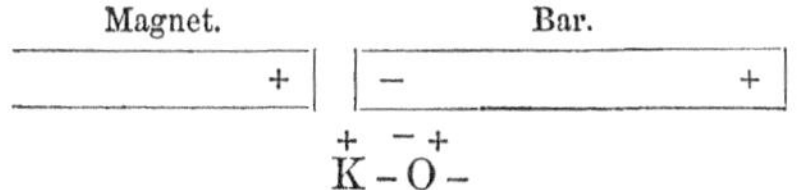

$$\overset{+}{K}-\overset{-+}{O}-$$

Furthermore, if we bring a lump of nickel in contact with the free pole of an iron bar, in the condition thus described, magnetic attraction will be developed in the mass of the nickel, a negative pole will be formed at the point of contact, and the lump will adhere. So, also, we may suppose that a similar effect is produced on the somewhat indifferent hydrogen-atom, which, added to K-O-, makes up the alkaline molecule—

$$\overset{+}{K}-\overset{-+}{O}-\overset{-}{H}.$$

Lastly, if we bring near the now loaded pole of our iron bar—to which we will assume there is attached as large a lump of nickel as it is capable of holding

—a lump of soft iron, the pole will drop the nickel and take the iron. In like manner, if we bring near our alkaline-molecule a radical, like NO_2, which has, by its own nature, or is capable of receiving by induction, a higher degree of negative polarity than the hydrogen-atom, then the molecule drops the hydrogen-atom and takes the radical.

Again, start with the same oxygen-atom with its two possible poles, and unite it by one of its bonds to a *negative* radical, it is evident that an opposite effect will be produced to that described in the last paragraph. The hydrogen-atom united to the remaining bond will now become by induction electro-positive, thus:

$$\overset{+}{H} - \overset{-+}{O} - (\overset{-}{N}O_2);$$

and, consequently, if we bring near the molecule a radical like K, which, by its nature, has a highly electro-positive polarity, the molecule will drop the hydrogen and take in its place the potassium atom. It is the preference for a negative radical in place of its hydrogen-atom which makes the first molecule alkaline, while it is a similar preference for a positive radical which renders the second molecule acid; and these preferences, as we now see, are manifestations of energy similar to those with which we are familiar in that well-known mode of polarity called magnetism.

Let me not, however, be understood to imply that the analogy here presented is perfect, or that it can be followed out into details; for this is far from being true. If chemism is, as it seems to be, a mode of polar action, it manifests characteristics which find their parallel in electrical rather than in magnetic phenomena. One instance of the failure of the analogy I have drawn we meet at once—and you have probably already detected it—in that important but small class of acids of which hydrochloric acid is the type. The molecules of these compounds consist of a single hydrogen-atom united to a highly-negative radical, and this hydrogen-atom has the same susceptibility of replacement by positive radicals, which is the essential characteristic of the acid hydrates we have been studying. These molecules contain no oxygen, and how, you may ask, can the theory of the constitution of acids and alkalies we have been expounding apply to them? The only answer we can give is, that they appear to present a simpler type of polarity, to which, though unlike magnetism, we have a parallel in the phenomena of electricity.

Take, for instance, the molecule of hydrochloric acid, HCl, the best example of its class. In this the chlorine-atom seems to have a *single* pole, which is strongly negative, and by its influence there appears to be induced an opposite pole, also *single*, in the atom of hydrogen. If, now, we bring near to this binary group an atom like Na, which either has by itself, or is capable of acquiring by induction, a higher degree of positive polarity than H, then the chlorine pole drops the H and takes the Na.

In the polar condition thus developed, the two opposite poles are on different atoms, and not only are the two atoms separable, but the positive or negative virtue appears to be inherent in the atom, and is transferred with it. A magnetic pole, on the contrary, is always associated with its opposite on the same mass of metal, and, if the mass is divided, two poles are found on each of the fragments, and so on indefinitely, however far the division may be carried. In the phenomena of statical electricity, however, we have a well-defined condition of polarity, to which the example of chemism we have been just discussing appears to be closely allied. If a pith-ball, electrified positively (or vitreously), is brought near a similar ball electrified negatively (or resinously), they attract each other, and the one becomes the pole of the other. If, now, the two are separated, each carries with it its electrical charge, and the peculiar virtue it has in consequence of that charge. But, though the two poles may thus be separated, and cease to have any relation to each other, yet they do not become isolated in any proper sense of that term, for each of the electrified bodies draws, by induction, an electrical charge, opposite to its own, to the extremity of the nearest conductor, and this charge becomes a new pole. An isolated pole is, in fact, a contradiction of terms. Polarity implies an opposition of relations, which involves two poles, and electrical polarity differs from magnetic polarity chiefly in the circumstance that the two poles are separate bodies. The magnetic poles are the ends of a polarized bar of iron, while the electrical poles are the boundaries of a mass of polarized dielectric, usually air, which intervenes between the oppositely electrified bodies; and every charge of electricity is just as closely associated with an opposite charge resting on some conductor beyond the insulating dielectric, as one magnetic pole accompanies the other.

Now, it is worthy of remark that this indissoluble association of opposite poles, which we must expect to find in chemical phenomena, if chemism is, as we suppose, a polar force, is actually manifested in a striking class of facts. The univalent atoms which, like those of chlorine or sodium, act as single poles, are never found isolated, but are always associated in a molecule with at least one other atom which forms the opposite pole of the molecular system, and, although the two poles of a molecule like HCl can be readily separated, the atoms do not remain isolated, but immediately form new associations, as in this very case, where the atoms of hydrogen pair off into molecules of hydrogen gas (H-H), and those of chlorine into molecules of chlorine gas (Cl-Cl), which are polar systems similar to those destroyed. On the other hand, bivalent atoms, like those of mercury or zinc, which have two poles, and may, therefore, constitute a complete polar system, each by itself, are sometimes found isolated, and form that class of molecules, previously described, in which the molecules consist of single atoms. The phenomena of quantivalence, also, which are such a characteristic feature of what we may now call chemical polarity, have their parallel in the phenomena of multiple poles, so familiar in magnetism, and may be caused by the same polar force acting through atoms of different shapes, and susceptibility to its influence; and the fact already referred to, that, in the variations of quantivalence, two bonds always appear or disappear at a time, is a strong confirmation of this theory; for, as has been said, one pole implies an opposite of equal strength, and the two must stand or fall together. It would be a further consequence of the theory that, although atoms of any even degree of quantivalence (artiads) might become isolated in molecules, those of an uneven degree (perissads) could not; and this also we find to be true so far as observation extends; but the number of elementary substances whose molecular weight has been directly determined is comparatively small, and those whose molecules are known to consist of single atoms, although all artiads, are only bivalent.

Returning now for a moment to the simple type of

polarity presented by the molecule H-Cl, let me call your attention to the fact that the polarity of the ordinary acid hydrates is but a modified form of the simpler type, and this will be obvious on comparing the symbol of hydrochloric acid with that of hypochlorous acid, from which it differs only by an atom of oxygen :

$$\underset{\text{Hydrochloric Acid.}}{\overset{+}{H}-\overset{-}{Cl}} \qquad \underset{\text{Hypochlorous Acid.}}{\overset{+}{H}-\overset{-+}{O}-\overset{-}{Cl}.}$$

You will notice that the atoms H and Cl are the poles of both systems, and that the oxygen-atom in the last is analogous to an armature between two magnetic poles, or, perhaps, more closely to a prime conductor between two oppositely-electrified balls :

$$\underset{H}{\oplus} \quad \underset{O}{(\overline{-\qquad+})} \quad \underset{Cl.}{\ominus}$$

Hypochlorous acid illustrates this relation more strikingly than nitric acid, our previous example of this class of compounds, but it is not nearly so stable a substance, and has never been obtained in a pure condition. Nitric acid differs from hypochlorous acid in containing a compound in place of a simple radical—

$$\underset{\text{Hypochlorous Acid.}}{\overset{+}{H}-\overset{-+}{O}-\overset{-}{Cl}.} \qquad \underset{\text{Nitric Acid.}}{\overset{+}{H}-\overset{-+}{O}-(\overset{-}{NO_2}).}$$

and the presence of compound radicals, often very complex, in the molecules of all the well-marked acids, necessarily increases the difficulty of interpreting their molecular structure, since the symbols may frequently be grouped in several ways without violating the principles of quantivalence. Our theory of the molecular structure of acid hydrates cannot, therefore, afford to waive the important evidence in its favor which has been obtained from recent investigations, and, as I am anxious to establish it on such a firm foundation that it may be taken as a basis in our further investigations of molecular structure, I must ask you to listen patiently to the few additional points I have to present.

The element carbon forms, with oxygen, besides the compound carbonic dioxide, which we have already studied, a second compound, called carbonic oxide, which has the symbol C=O. In this molecule two of the bonds of the carbon-atom are unemployed, or, rather, neutralized by their mutual attraction. Hence, these molecules are very much in the same condition as the atoms of mercury or zinc, when acting as molecules, and, like them, the molecule CO can enter into direct combination, as a bivalent radical. Striking instances of such combination are the formation of phosgene gas by the direct union of carbonic oxide with chlorine gas, under the influence of sunlight, and the burning of carbonic oxide, when the same molecules unite with an additional atom of oxygen to form carbonic dioxide:

$$\underset{\text{Carbonic Oxide.}}{CO} + \underset{\text{Chlorine Gas.}}{Cl-Cl} = \underset{\text{Phosgene Gas.}}{COCl_2}, \text{ or } Cl-\overset{\overset{\displaystyle O}{\|}}{C}-Cl;$$

$$\underset{\text{Carbonic Oxide.}}{2CO} + \underset{\text{Oxygen Gas.}}{O=O} = \underset{\text{Carbonic Dioxide.}}{2CO_2} \text{ or } O=C=O.$$

Now, if potassic hydrate, K-O-H, is gently heated in an atmosphere of carbonic oxide, a slow but regular absorption of the gas takes place, and the potassium salt of a well-known acid, called formic acid, is the result, and, from a mixture of this salt with sulphuric acid, we can readily distill off the acid itself. Formic acid being volatile, we can determine with certainty its molecular weight, and, since an accurate analysis is also possible, there is no doubt whatever that the symbol H_2O_2C expresses the exact composition of its molecule. But how are these atoms arranged? As data for solving this problem, we have, in the first place, the known quantivalence of the several atoms, and, in the second place, a knowledge of the fact, acquired in studying the phenomena of combustion, that, if, in the reaction by which formic acid was produced, the two atoms of the radical CO had been parted, an enormous absorption of heat must have attended the chemical change. But no such thermal effect, nor any of the phenomena, which would naturally accompany it, have been noticed, and we therefore feel justified in concluding that the radical CO exists as such in formic acid, as the direct absorption of the gas by caustic potash would seem to indicate. The only question that remains is, how the other atoms are grouped around this radical, and the quantivalence of the atoms permits but one mode of grouping, as follows:

$$\begin{array}{c} \quad\quad\;\; O \\ \quad\quad\;\; \| \\ H-O-C-H. \end{array}$$

In this molecule there are two atoms of hydrogen, one united directly to the carbon-nucleus, the other also united to the same radical, but only indirectly through the atom of oxygen which intervenes. Now, are both of these hydrogen-atoms equally susceptible of replacement? We find not. If we neutralize the acid by potassic hydrate, we obtain the same potassium salt which was formed by the direct union of the alkali with carbonic oxide, and analysis shows that this salt contains just one-half as much hydrogen as the acid from which it was formed, and, by no metathetical reaction whatever can we succeed in replacing the remaining atom.

Evidently, then, the two atoms stand in very different relations to the molecule; but which was the one replaced? As to this point, we have the most conclusive and abundant evidence. We need call, however, but a single class of witnesses. Formic acid is the first of a series of volatile acids, and the molecules of the successive compounds which form the steps of this series differ from each other by the common difference CH_2. The second member of the series is acetic acid, which, in a diluted condition, is used as a condiment with our food under the name of vinegar. The composition of pure acetic acid is represented by the symbol $H_4O_2C_2$, and the molecule of this acid, therefore, contains four atoms of hydrogen. But of these only one is replaceable —as in formic acid—and the same is true of all the acids of this class, although the molecules of the last member of the series contains no less than sixty hydrogen-atoms. Moreover, acetic acid—like formic acid—contains two atoms of oxygen, and two corresponding atoms—and only two—appear in the molecules of all the other members of the same series. Add now the further fact, which will be illustrated more fully hereafter, that several of the compounds in the series have been prepared from formic acid by processes which show that, if the radical

$$H-O-\overset{\overset{\large O}{\|}}{C}-$$

exists in the molecule of this, the first member of the series, it must also form a part of the molecules of all the other members, and you will be prepared, I think, to answer the question proposed above. The facts stated may be almost said to prove that in all these molecules one, and only one, atom of hydrogen is united to the radical by an atom of oxygen, and this must be the single atom which in all these compounds is susceptible of replacement. We may, therefore, write the symbol of formic acid—

$$\overset{+}{H}-\overset{-+}{O}-(\overset{-}{C}OH),$$

and regard the molecule as having a polar condition like that we attributed to the molecule of nitric acid.

Here, then, is a well-marked acid, in regard to the structure of whose molecule there can be no reasonable doubt, and the conclusion we have reached in regard to it harmonizes completely with that we had previously formed in regard to the structure of the molecule of nitric acid on wholly different grounds. Such a concurrence of testimony gives us great confidence in the theory we have advanced in regard to the constitution of this class of substances, and we may certainly accept it as a trustworthy guide in the further prosecution of our study.

It will not, of course, be for a moment inferred that we regard the argument now concluded as demonstrative. We have been advocating what we have expressly called a theory, and all we claim is that the evidence advanced is sufficiently conclusive to render the theory credible, and that the theory is of great value, both by giving us a more comprehensive grasp of the facts with which we have to deal, and by helping us to associate the supersensuous phases of molecular action with the visible phenomena of magnetism and electricity.

Having then stated, as fully as the circumstances will permit, the evidence on which our theory of the constitution of acids and alkalies rests, in the case of a few of the simpler of these compounds, I must, as regards the molecular structure of the more complex compounds of the same type, content myself with merely stating results, only premising that the conclusions rest on evidence similar to that already adduced.

Beginning with the series of volatile acids, of which formic and acetic acids are members, let me first call your attention to the following symbols, which, as we believe, represent the molecular structure of these bodies :

Formic acid............

$$\begin{array}{ccccccc} & & & & \text{O} & & \\ & & & & \| & & \\ \text{H} & - & \text{O} & - & (\text{C} & - & \text{H}) \end{array}$$

Acetic acid............

$$\begin{array}{ccccccccc} & & & & \text{O} & & \text{H} & & \\ & & & & \| & & | & & \\ \text{H} & - & \text{O} & - & (\text{C} & - & \text{C} & - & \text{H}) \\ & & & & & & | & & \\ & & & & & & \text{H} & & \end{array}$$

Propionic acid..........

$$\begin{array}{ccccccccccc} & & & & \text{O} & & \text{H} & & \text{H} & & \\ & & & & \| & & | & & | & & \\ \text{H} & - & \text{O} & - & (\text{C} & - & \text{C} & - & \text{C} & - & \text{H}) \\ & & & & & & | & & | & & \\ & & & & & & \text{H} & & \text{H} & & \end{array}$$

Normal butyric acid ....

$$\begin{array}{ccccccccccccc} & & & & \text{O} & & \text{H} & & \text{H} & & \text{H} & & \\ & & & & \| & & | & & | & & | & & \\ \text{H} & - & \text{O} & - & (\text{C} & - & \text{C} & - & \text{C} & - & \text{C} & - & \text{H}) \\ & & & & & & | & & | & & | & & \\ & & & & & & \text{H} & & \text{H} & & \text{H} & & \end{array}$$

Normal valeric acid.....

$$\begin{array}{ccccccccccccccc} & & & & \text{O} & & \text{H} & & \text{H} & & \text{H} & & \text{H} & & \\ & & & & \| & & | & & | & & | & & | & & \\ \text{H} & - & \text{O} & - & (\text{C} & - & \text{C} & - & \text{C} & - & \text{C} & - & \text{C} & - & \text{H}) \\ & & & & & & | & & | & & | & & | & & \\ & & & & & & \text{H} & & \text{H} & & \text{H} & & \text{H} & & \end{array}$$

All the above compounds have been thoroughly investigated, and all the symbols given above rest on as good evidence as the first. All these compounds have the same general structure, and the same system of polarity, as the simpler hydrates, and they may be regarded as derived from formic acid by successive substitutions of

$$\begin{array}{cccc} & \text{H} & & \\ & | & & \\ - & \text{C} & - & \text{H} \\ & | & & \\ & \text{H} & & \end{array}$$

for the final hydrogen-atom of the negative radical. Lastly, notice the binary group, H-O-, which plays such an important part in these and all similar molecules. This group of atoms, or radicals, has been named *hy-*

droxyl, and, for the future, we shall find it convenient to employ this term.

In all the examples thus far cited, in illustration of our theory of the molecular structure of acid and alkaline hydrates, the molecule has contained but one hydroxyl (HO) group, and therefore but one replaceable hydrogen-atom. Such hydrates are said to be monoatomic. While, however, the univalent radicals, which these compounds all contain, can only bind one hydroxyl group, a bivalent radical may be associated with two such groups, a trivalent radical with three, and so on. In the resulting compound there will be as many replaceable atoms of hydrogen as there are hydroxyl groups united to the radical, and the number of these replaceable atoms measures what is called the atomicity of the compound. We are now prepared to define also the term *hydrate*, that we have so frequently used in this lecture to designate the class of compounds to which all the alkalies and most of the acids belong. *A hydrate is, simply, a compound of hydroxyl, and is monatomic, diatomic, triatomic, etc., according as it contains one, two, three, or more hydroxyl groups.* Let me illustrate this important principle by a few examples of hydrates of multivalent radicals, beginning with those in which the radical is bivalent.

At the boiling-point, metallic magnesium slowly decomposes water, liberating hydrogen gas—

$$\underset{\text{Water.}}{2H_2O} + \underset{\text{Magnesium.}}{Mg} = \underset{\text{Magnesic Hydrate.}}{MgO_2H_2} + \underset{\text{Hydrogen Gas.}}{H\text{-}H}$$

In this reaction the bivalent atom of magnesium binds together two molecules of water to form a molecule of magnesic hydrate, whose structure may be represented:

$$\underset{\text{Magnesic Hydrate.}}{H-O-Mg-O-H.}$$

The molecule of common slacked lime, calcic hydrate, has a similar structure:

$$\mathrm{H-O-Ca-O-H.}$$

Calcic Hydrate.

These two hydrates are both alkaline, but there are corresponding acid hydrates, among which are numbered the two very important chemical agents called sulphuric and oxalic acids, whose molecules are supposed to have the structure indicated by our diagrams:

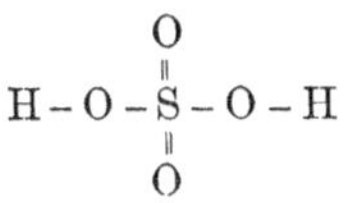

$$\begin{array}{c} \mathrm{O} \\ \| \\ \mathrm{H-O-S-O-H} \\ \| \\ \mathrm{O} \end{array} \qquad \begin{array}{c} \mathrm{O\ \ O} \\ \|\ \ \| \\ \mathrm{H-O-C-C-O-H} \end{array}$$

Sulphuric Acid. 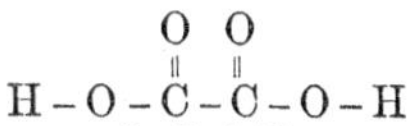 Oxalic Acid.

Compounds like the last four are said to be diatomic; for there are in each case two hydroxyl groups, and therefore two easily-replaceable atoms of hydrogen, and this is shown, in the case of the acids, by the fact that, when wholly or one-half neutralized with caustic soda or potash, they give two different salts, in one of which the whole, and in the other only one-half, of the hydrogen of the acid is replaced. Thus, we have—

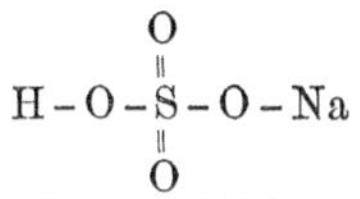

$$\begin{array}{c} \mathrm{O} \\ \| \\ \mathrm{H-O-S-O-Na} \\ \| \\ \mathrm{O} \end{array} \qquad \begin{array}{c} \mathrm{O} \\ \| \\ \mathrm{Na-O-S-O-Na} \\ \| \\ \mathrm{O} \end{array}$$

Hydrosodic Sulphate. 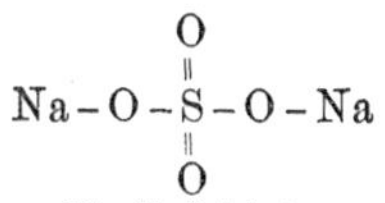 Disodic Sulphate.

So also—

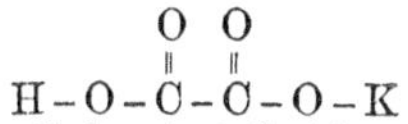

$$\begin{array}{c} \mathrm{O\ \ O} \\ \|\ \ \| \\ \mathrm{H-O-C-C-O-K} \end{array} \qquad \begin{array}{c} \mathrm{O\ \ O} \\ \|\ \ \| \\ \mathrm{K-O-C-C-O-K} \end{array}$$

Hydropotassic Oxalate. 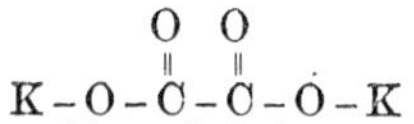 Dipotassic Oxalate.

If, however, we neutralize these dibasic acids with magnesic or calcic hydrates, we can obtain but one product, because the bivalent atoms Mg and Ca replace the two hydrogen atoms at once. The salts thus obtained have the symbols:

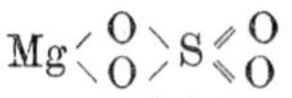

$$Mg\langle{}^{O}_{O}\rangle S\langle{}^{O}_{O}$$
Magnesic Sulphate.

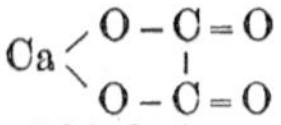

$$Ca\langle\begin{matrix} O-C=O \\ \quad | \\ O-C=O \end{matrix}$$
Calcic Oxalate.

It may, perhaps, avoid some confusion to repeat here the remark already made, that the position or grouping of the symbols on the diagram is wholly arbitrary beyond the relations which the dashes indicate.

Pass next to hydrates which contain three hydroxyl groups, and are, therefore, said to be triatomic. Of these we shall only cite two examples :

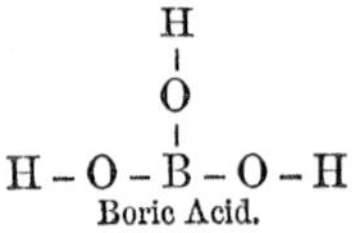

$$\begin{matrix} & & H & & \\ & & | & & \\ & & O & & \\ & & | & & \\ H-O & - & B & - & O-H \end{matrix}$$
Boric Acid.

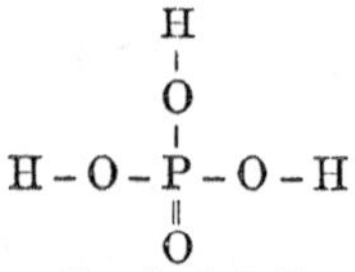

$$\begin{matrix} & & H & & \\ & & | & & \\ & & O & & \\ & & | & & \\ H-O & - & P & - & O-H \\ & & \| & & \\ & & O & & \end{matrix}$$
Phosphoric Acid.

The triatomic character of phosphoric acid is shown by the fact that it can be neutralized by caustic soda in three successive stages, and gives three compounds, one of which contains no hydrogen, and the others respectively one-third and two-thirds as much as in the corresponding quantity of the acid. The names and symbols of these salts are as follows :

$Na_3{\equiv}O_3{\equiv}PO$	$H,Na_2{\equiv}O_3{\equiv}PO$	$H_2,Na{\equiv}O_3{\equiv}PO$
Trisodic Phosphate.	Hydrodisodic Phosphate.	Dihydrosodic Phosphate.

This abbreviated form of notation can be easily understood, and requires no further explanation. It saves space in printing, and gives all the data required for constructing the graphic symbols.

Of hydrates containing four hydroxyl groups, therefore, tetratomic, the most familiar is silicic hydrate—

$$\begin{matrix} H-O \\ H-O \end{matrix}\rangle Si\langle\begin{matrix} O-H \\ O-H \end{matrix}$$

but this substance is very unstable, and hitherto it has

been impossible to prepare it of constant composition. The instability is due to a cause which is inherent in many of the more complex molecular structures. Wherever there is a tendency in the atoms to group themselves, so as to better satisfy their mutual affinities, a slight cause is sufficient to destroy the balance of forces on which the existence of the molecule depends, and the structure breaks up into simpler parts. The explosion of nitro-glycerine was a conspicuous example of this principle, and we have, in these complex hydrates, another illustration of the same. It is evident, from the very great amount of heat evolved in the direct union of oxygen and hydrogen gases, that the molecules of water are in a condition of great stability, and the hydrogen and oxygen atoms, which are associated in such numbers in the molecules of the more complex hydrates, are constantly tending to this condition of more stable equilibrium. Indeed, these compounds give off water so readily, either spontaneously or at the slightest elevation of temperature, that they were formerly supposed to contain water, as such, and hence the name *hydrates* (from ὕδωρ, *water*), which has been retained in our modern nomenclature, although with a modified meaning.

Since the number of oxygen and hydrogen atoms in the several hydroxyl groups united to the radical of a hydrate must necessarily be the same, it follows that the formation of every molecule of water must be attended with the liberation of an atom of oxygen, and, when a hydrate breaks up, these atoms frequently unite with the radical to form compound radicals of lower quantivalence. Thus we have formed from the normal silicic hydrate, by the elimination of successive molecules of water, the following products:

```
       H               H
       |               |
       O               O
       |               |
H - O - Si - O - H     Si = O         O = Si = O
       |               |
       O               O
       |               |
       H               H
 Normal Hydrate.   First Anhydride.[1]   Second Anhydride.
```

The atoms of oxygen liberated as just described may also bind together several atoms of silicon, and thus give rise to still more complex groups, such as—

```
      H      H                      H      H      H
      |      |                      |      |      |
      O      O                      O      O      O
      |      |                      |      |      |
H - O - Si - O - Si - O - H   H - O - Si - O - Si - O - Si - O - H
      |      |                      |      |      |
      O      O                      O      O      O
      |      |                      |      |      |
      H      H                      H      H      H

      H      H                H                   H
      |      |                |                   |
      O      O                O                   O
      |  /O\ |                |  /O\     /O\      |
      Si     Si               Si     Si      Si
      |  \O/ |                |  \O/     \O/      |
      O      O                O                   O
      |      |                |                   |
      H      H                H                   H
```

These compounds may be regarded as formed by the coalescing of two or more molecules of the normal hydrate, and the elimination from these combined molecules of successive molecules of water as before. The following table will illustrate what is meant:

$H_4 O_4 Si$	$2(H_4 O_4 Si)$	$3(H_4 O_4 Si)$
$H_2 O_2 SiO$	$H_6 O_6 Si_2O$	$H_{10} O_{10} Si_3O$
SiO_2	$H_4 O_4 Si_2O_2$	$H_8 O_8 Si_3O_2$
	$H_2 O_2 Si_2O_3$	$H_6 O_6 Si_3O_3$
	$2SiO_2$	$H_4 O_4 Si_3O_4$
		$H_2 O_2 Si_3O_5$
		$3SiO_2$

[1] A compound derived from a hydrate by the elimination of water is called an anhydride.

The table might be extended indefinitely. It is true that not every member of these series is even theoretically a possible compound; but, by attempting to write the symbols in the more graphic form, those cases in which the atoms cannot be grouped in a single molecule will be readily distinguished.

We have in this glass a solution of sodic silicate, which is commonly called soluble glass. On adding to the solution some muriatic acid, you notice that there is at once formed a white, bulky, gelatinous mass. This is supposed to be the normal silicic hydrate, but, when we attempt to wash and dry the substance for the purpose of analysis, it begins to lose water, and we have found it impossible to arrest the change at any definite point. In the process of drying, the various hydrates, whose symbols we have given, are probably produced, but only as passing phases of the dehydration, and these symbols would be wholly ideal were it not that, on replacing the hydrogen-atoms by metallic radicals, we obtain products of great stability. The compounds to which I refer are the mineral silicates that form so large a part of the minerals and rocks of the globe. The two following well-known, although not abundant, minerals correspond, for example, to the normal hydrate and its first anhydride respectively:

$$Mg\langle{}^{O}_{O}\rangle Si\langle{}^{O}_{O}\rangle Mg \qquad\qquad Ca\langle{}^{O}_{O}\rangle Si=O$$

Magnesia Chrysolite. Wollastonite.

and the symbols show that the molecular structures we have described above are realized in these natural products if not in the hydrates. The molecular structure of some of our most common minerals, such as feldspar and garnet, corresponds to that of some of the most complex hydrates, with radicals consisting of several silicon-atoms; but, we shall understand better the manner in which these highly-complex molecules are built up, after we have become acquainted with a remarkable hexatomic hydrate, whose well-marked sexivalent radical plays a very important part in their structure.

No definite pentatomic hydrate is known, but of hexatomic hydrates there are several noteworthy examples. The one referred to in the last paragraph is the hydrate of aluminum. The normal hydrate of this element, and the several anhydrides which may be formed from it by the elimination of successive molecules of water, are all well-defined mineral substances. The following table shows the relations of these compounds to each other, and also to certain other mineral substances in which the hydrogen-atoms have been replaced:

$Al_2{}^{vi}O_6{}^{vi}H_6$	$O{=}Al_2{\equiv}O_4{\equiv}H_4$	$O_2{\equiv}Al_2{=}O_2{=}H_2$	$O_3{}^{vi}Al_2$
Gibbsite.	Beauxite.	Diaspore.	Corundum.
	$O{=}Al_2{\equiv}O_4{\equiv}Si$	$O_2{\equiv}Al_2{=}O_2{=}G$	
	Andalusite.	Chrysoberyl.	

It would be interesting to represent in a graphic form these molecules, but I can leave this to your own study, and close my illustrations of the subject with two or three examples of the very highly-complex molecular structures which the salts of aluminum present, and in which the mode of atomic grouping is less obvious:

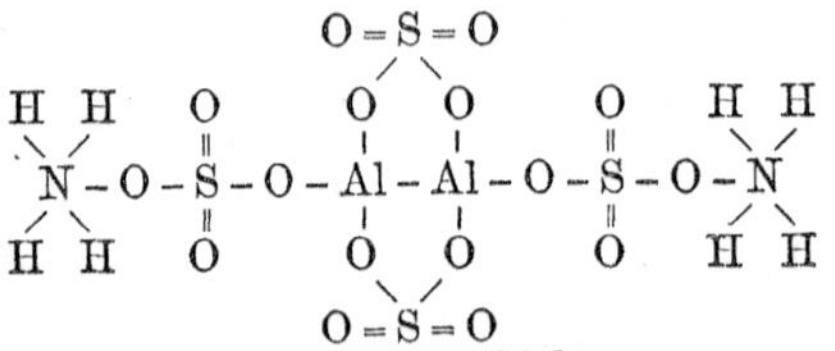

```
                      O = S = O
                         / \
 H  H     O        O     O       O     H  H
  \ /     ‖        |     |       ‖      \ /
   N - O - S - O - Al - Al - O - S - O - N
  / \     ‖        |     |       ‖      / \
 H  H     O        O     O       O     H  H
                         \ /
                      O = S = O
```

Ammonia Alum (dried.)

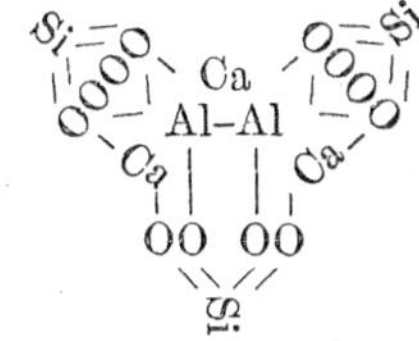

Garnet (Lime Alumina).

```
                 O
                / \
             Si - O - Si
            /  \     /  \
           O    O   O    O
          /      \ /      \
K - O - Si - O - Al - Al - O - Si - O - K
          \      /  \      /
           O    O    O    O
            \  /      \  /
             Si - O - Si
                \   /
                  O
```

Feldspar (Orthoclase).

In arranging these symbols for our diagrams, we naturally seek a symmetrical disposition; but it must not be forgotten that every thing beyond the number of atomic bonds, and the relative position which the dashes indicate, is purely arbitrary.

I have dwelt at this length on the theory of the acid and alkaline hydrates, because it is just here that the distinction between the new-school and the old-school chemistry chiefly appears. The dualistic theory, which originated with Lavoisier, and was extended and illustrated by Berzelius, was based on the very class of facts we have been studying in the two preceding lectures of this course. At the time of Berzelius, the elements, the acids, the alkalies, or bases, and the large class of compounds called salts, made up very nearly the whole of chemistry, and, of the facts then known, the dualistic theory gave a satisfactory explanation. It was the natural outgrowth of the discovery of oxygen gas, that universally-diffused element with which all other elementary substances combine, and of whose compounds almost the whole of terrestrial Nature consists. Lavoisier inferred that oxygen must be the chemical centre in the scheme of Nature, and he therefore made its compounds the basis of a new classification, which, subsequently, Berzelius greatly systematized and improved. In this classification the compounds of the elements with oxygen were divided into two classes: Those which, when dissolved in water—combined with it we should now say—gave an acid reaction, were called acids; while those which, under the same circumstances, gave an alkaline reaction, were called bases. It was known then, as well as now, that these reactions could not be obtained without the presence of water, and that the larger part of the oxides, being insoluble in water, do not give the reactions at all; but, then it was supposed that the water acted only through virtue of its solvent power, that some other solvent would do as well, and that the insoluble oxides would give the same reactions if only an appropriate solvent could be found. Hence, these insoluble oxides were classed with the acids or bases, according as they combined most readily with bases or acids respectively. The insoluble SiO_2 combined with soda, like the soluble SO_3, and hence was classed with it as an acid. So the insoluble FeO combined with sulphuric acid, like the soluble CaO, and hence was classed with the last as a base. Again, the neutralizing of an acid by an alkali had all the appearance of direct combination, and, in all these processes, the acid oxide was assumed to unite with the metallic, or basic, oxide to form what was called a salt. The presence of the water, and the fact that it facilitated the chemical change, were not ignored, but, as before, it was supposed to act in virtue of its solvent power, and a sufficient number of cases were known where the same compounds could be obtained with and without the aid of water to render this opinion not improbable. Take a single example: Phosphate of lime may be made in two ways: first, by adding to a solution of lime in water a solution of phosphoric acid:

$$(3Ca{=}O_2{=}H_2 + 2H_3{\equiv}O_3{\equiv}PO + Aq.) = Ca_3{}^{vi}O_6{}^{vi}(PO)_2 + (6H\text{-}O\text{-}H + Aq.).$$

Secondly, by uniting lime, the oxide of the metal calcium, directly to P_2O_5, the oxide obtained by burning phosphorus (page 185):

$$3CaO + P_2O_5 = 3CaO,P_2O_5, \quad \text{or} \quad Ca_3{}^{vi}O_6{}^{vi}(PO)_2.$$

In the last reaction there is no water present, and the first reaction was formerly supposed to be a case of similar direct union between CaO and P_2O_5, the only difference being that the two oxides were in solution:

$$3(CaO,H_2O) + 3H_2O,P_2O_5 = 3CaO,P_2O_5 + 6H_2O.$$

Accordingly, it was customary to write the symbols as in this last reaction, separating the acid from the basic oxide by a comma. Here are a few other examples:

CaO,SO_3	FeO,SO_3	ZnO,N_2O_5.[1]
Sulphate of Lime.	Sulphate of Iron.	Nitrate of Zinc.

As expounded and illustrated by Berzelius, the dualistic theory had the charm of great simplicity, and was greatly strengthened by the electro-chemical facts which he brought forward in its support. The division of the elementary substances into electro-positive and electro-negative elements corresponded very closely to

[1] To avoid confusion, all our symbols stand for the new atomic weights, and this must be remembered in comparing these formulæ with those in the old books.

the distinction between metals and metalloids. Bases were compounds of electro-positive elements with oxygen; and acids, on the other hand, the oxides of electro-negative elements. Again, among these binary compounds the basic oxides were electro-positive, and the acid oxides electro-negative. Moreover, the wider apart in their electrical relations, the stronger was seen to be the tendency of both the elements and of their oxides to combine, and, just as the metals united to metalloids, so bases united with acids. Thus was formed the class of ternary compounds, called, as above, salts.[1] Among these, also, could be distinguished a similar opposition of relations, although less marked, to that between bases and acids, and, from the union of two salts, resulted the class of quaternary compounds, or double salts. In this way the theory advanced from elementary substances to the most complex compounds through the successive gradations of binaries, ternaries, and quaternaries; the elements or compounds only combining with substances of the same order, two and two together, like two magnetic poles, or two electrified bodies.

This dualistic theory was certainly a most admirable system, and served the purposes of a rapidly-grow-

[1] The word *salt* was used in chemistry very early to describe any saline substance resembling externally common salt; but, under the dualistic system, the term came to be applied to that class of compounds which were supposed to be formed by the union of basic and acid oxides, as described above. Absurdly enough, however, common salt was thus ruled out of the very class of compounds of which it had previously been regarded as the type, and Berzelius, in his electro-chemical classification, made a distinct family of those substances which resemble common salt in their chemical composition, and called it the *haloids*. But this name —*bodies resembling salt*—only rendered the anomaly the more glaring, and it was always a blemish on the dualistic system. In the modern chemistry, the word *salt*, although still used as a descriptive name, has no technical meaning.

ing science for more than half a century. We now feel assured that the old theory undervalued essential circumstances, and misinterpreted important facts. We maintain that hydrogen is an essential, not an accidental constituent of all acids and all alkalies, and that, when the alkali is neutralized by the acid, the reaction consists in the replacement of this hydrogen, and not in the direct union of two oxides. Nevertheless, given the old facts, the old theory was logical and consistent, and it is no longer tenable, not because the old facts have changed, but simply because a whole new order of facts has been discovered by which the old facts must be interpreted. During the last twenty-five years there has been discovered a great mass of truths, connected chiefly with the compounds of carbon, in what was formerly called the domain of organic chemistry, and this is to-day the most prominent and attractive portion of our science. Moreover, the law of Avogadro and the doctrine of quantivalence are two new principles which our modern science has added to the old chemistry, and these principles have supplanted the dualistic theory. Let us not, however, undervalue the old theory. It was an important stage in the progress of science, and a noble product of human thought. Theories are means, not ends; but they are the appointed means by which man may raise himself above the low level of merely sensuous knowledge to heights where his intellectual eye ranges over a boundless prospect which it is the special privilege of the student to behold. What though his vision be not always clear, and his imagination fill the twilight with deceptive shapes which vanish as the light of knowledge dawns; yet, to have enjoyed the intellectual elevation, is reward enough for all his devotion and all his toil.

LECTURE XIII.

ISOMERISM, AND THE SYNTHESIS OF ORGANIC COMPOUNDS.

Having, in the previous lectures of this course, made you familiar with the conception that the molecules of every substance have a definite atomic structure, which is a legitimate object of scientific investigation, I endeavored in my last lecture to illustrate, by numerous examples, the mode now generally employed in chemistry of exhibiting this structure by means of what are called structural formulæ, and, during the whole course of these lectures, it has been a chief object to develop the fundamental principles on which these formulæ are based, in order that, having reached this stage, you might be able to see for yourselves that they were legitimately deduced from the facts of observation. I have freely admitted that they were the expression of theoretical conceptions which we could not for a moment believe were realized in Nature in the concrete forms, which our diagrams embody. But I have claimed that they were at present our only mode of representing to the mind a large and important class of facts, and were to be valued as the first glimpses of some great, general truth, toward which they direct our investigation. Theories are the only lights with which we can penetrate the obscurity of the unknown, and they are to be valued just so far as they illuminate our path. This ability to lead investigation is the only true test of any theory, and it will be my object in this my last lecture to show that the modern chemical theory of molecular structure has a claim to be regarded as one of the most valuable aids to discovery which science has ever received.

The illustrations of molecular structure thus far studied have been mostly taken from those classes of compounds long known in chemistry under the names of acids, bases, and salts, and they were selected because it was with such substances that the old theory had almost exclusively to deal, and they were therefore the best adapted to illustrate the differences between the new and the old chemistry. But, as I have already said, the strongest evidence in favor of the new theory is to be obtained from a class of substances about which the old chemistry knew almost absolutely nothing, and whose number has been enormously increased during the past twenty-five years. Indeed, the modern theory is so completely the outgrowth of new discoveries that, given alone the old facts, the question between the old and the new theories would be at least of doubtful issue, even if the new could ever have been conceived. The class of substances to which I refer are the compounds of the elementary substance carbon. The number of known compounds of this one element is far greater than that of all the other elements besides, and these compounds exhibit a great diversity in their molecular structure, which is often highly complex. As a rule they consist of a very few chemical elements (besides carbon, only hydrogen, oxygen, and nitrogen), but the number of atoms united in a single molecule may be very large, sometimes even exceeding one hundred. Carbon is peculiarly the element of the organic world, for, leaving out of view the great mass of water which living beings always contain, organized material consists almost exclusively of carbonaceous compounds. Hence these substances, with the exception of a few of the simplest, were formerly called organic compounds, and in works on chemistry they are usually studied together under the head of organic chemistry. It was formerly supposed that the great complexity of these substances was sustained by what was called the vital principle; but, although the cause which determines the growth of organized beings is still a perfect mystery, we now know that the materials of which they consist are subject to the same laws as mineral matter, and the complexity may be traced to the peculiar qualities of carbon. In like manner the notion that these so-called organic substances owed their origin to some mysterious energy, which overruled the ordinary laws of chemical action, for a long time precluded from the mind of the chemist even the idea that they could be formed in the laboratory by purely chemical processes; so that, although the analysis of these compounds was easily effected, the synthesis was thought impossible. But within a few years we have succeeded in preparing artificially a very large number of what were formerly supposed to be exclusively organic products; and not only this, but the processes we have discovered are of such general application that we now feel we have the same command over the synthesis of organic, as of mineral substances. The chemist has never succeeded in forming a single organic cell, and the whole process of its growth and development is entirely beyond the range of his knowledge; but he has every reason to expect that, in the no distant future, he will be able to prepare, in his laboratory, both the material of which that cell is fashioned, and the various products with which it becomes filled during life.

The number of elements which enter into the composition of organic compounds being so restricted, it is evident that the immense variety of qualities which they present cannot be referred solely to the influence of the simple radicals which they contain.[1] Moreover, there appears among these organie substances a most remarkable phenomenon, which, although not unknown in the mineral kingdom, is peculiarly characteristic of these complex compounds of carbon. We are acquainted with a large number of cases of two or more wholly different substances having exactly the same composition and the same vapor density. Here, for example, are two such substances:

The first, butyric acid, is an oily liquid with whose smell we are only too familiar, since, when formed in rancid butter, it imparts to this article of our food its peculiarly offensive odor. But, though, as the odor shows, it must slowly volatilize at the ordinary temperature, it does not boil lower than 156° C., and does not easily inflame. Further, as its name denotes, it has the qualities of an acid, reddening litmus-paper, and causing an effervescence with alkaline carbonates.

Utterly different from this offensive acid is the second substance, which we call acetic ether, a very limpid liquid, with a pleasant, fruity smell, highly volatile, boiling at 74° and inflaming with the greatest ease. Notice, also, that it does not in the least affect the colors of these sensitive vegetable dyes.

Yet, butyric acid and acetic ether have exactly the

[1] Compare pages 243 and 262.

same composition, and the same vapor density. The results both of actual chemical analysis and of the determination of vapor density are given in this diagram, and the figures obtained in the two cases do not differ more than we should expect the results of different analyses of the same substances to differ; for it must be remembered that, in such experimental work, we can only attain a certain degree of accuracy, and that we may disregard all variations which are within the limit of probable error:

Analyses of Isomeric Compounds.

By Grünzweig.		By Liebig.	
Butyric Acid—		Acetic Ether—	
Carbon........	54.51	Carbon.......	54.47
Hydrogen......	9.26	Hydrogen.....	9.67
Oxygen........	36.23	Oxygen.......	35.86
	100.00		100.00
By Cahours.		By Boullay and Dumas.	
Sp. Gr............	44.3	Sp. Gr...........	44.1
Molec. weight.....	88.0	Molec. weight....	88.0

If, now, from these experimental results, we come to calculate the symbols of the two substances, according to the method I have so fully described, we shall obtain in both cases precisely the same formula, $C_4H_8O_2$, and it must, therefore, be that the molecules of these two substances contain the same number of atoms of the same three elements, carbon, hydrogen, and oxygen. Here, then, we come face to face with a most remarkable fact. For, to affirm no more than can be absolutely demonstrated, this pleasant odor of apples and this disgusting smell of rancid butter come from substances consisting of the same elements united in the same proportions. What, then, can be the cause of the difference? We cannot allow such a fundamental fact as this to pass unchallenged. It is evident that there is an all-important condition which has escaped our elementary analysis. The circumstances demand investigation, and it would be a disgrace to our science not to attempt to answer the question. Can you wonder, then, that, for the past ten years, a great part of the intellectual force of the chemists of the world has been applied to the problem, and in this course of lectures I have been endeavoring to present to you the result they have reached. The answer they have obtained is, that the difference of qualities depends on molecular structure, and that the same atoms arranged in a different order may form molecules of different substances having wholly different qualities. But they have gained more than this general result.

These *isomeric compounds*, as we call them, when acted on by chemical agents, break up in very different ways, and, by studying the resulting reactions, we are frequently able to infer that certain groups of atoms (or compound radicals) are present in the compounds, because we know that they exist in the products which these compounds respectively yield; our knowledge of the structure of these very radicals probably depending on yet other reactions, by which they again may be resolved into still simpler groups.

Thus, for example, if we act on acetic ether with potassic hydrate, we obtain two products, potassic acetate and common alcohol. Now, we know that alcohol has the symbol C_2H_5-O-H and contains the radical C_2H_5, which we call ethyl. Further, we know that potassic acetate has the symbol K-O-(C_2H_3O) and contains the radical C_2H_3O, which we call acetyl. Hence we infer that the ether contains both of these groups, and that its symbol must be C_2H_5-O-C_2H_3O. The reaction obtained with potassic hydrate is, then, seen to consist in a simple metathesis between K and C_2H_5.

$$\left.\begin{array}{c} C_2H_5\text{-}O\text{-}C_2H_3O \\ \text{Acetic Ether.} \\ K\text{-}O\text{-}H \\ \text{Potassic Hydrate.} \end{array}\right\} = \left\{\begin{array}{c} K\text{-}O\text{-}C_2H_3O \\ \text{Potassic Acetate.} \\ C_2H_5\text{-}O\text{-}H \\ \text{Alcohol.} \end{array}\right.$$

Passing next to the radical ethyl C_2H_5, we can show that it may be formed in a compound which contains the radical CH_3, called methyl, by substituting for one of the hydrogen-atoms of this radical another group of the atoms CH_3, thus:

$$\left.\begin{array}{c} \begin{array}{ccccc} & & H & & \\ & & | & & \\ H & - & C & - & X \\ & & | & & \\ & & H & & \end{array} \\ \text{First Methyl Compound.} \\ \\ \begin{array}{ccccc} & & H & & \\ & & | & & \\ H & - & C & - & Y \\ & & | & & \\ & & H & & \end{array} \\ \text{Second Methyl Compound.} \end{array}\right\} = \left\{\begin{array}{c} \begin{array}{ccccccc} & & H & & H & & \\ & & | & & | & & \\ H & - & C & - & C & - & X \\ & & | & & | & & \\ & & H & & H & & \end{array} \\ \text{Ethyl Compound.} \\ \\ H - Y \\ \\ \text{Hydrogen Compound.} \end{array}\right.$$

In this assumed reaction the terminal hydrogen-atom of the first methyl compound changes place with the methyl radical of the second, thus producing the compounds in the second column. Such a reaction can actually be produced with a variety of substances, and these symbols may be supposed to stand for any of the substances between which the reaction is possible. We use X and Y, instead of writing the symbols of definite compounds, in order to confine the attention to the change which takes place in the radical alone.

In reactions of this kind we form the radical ethyl in such a way as to leave no doubt whatever in regard to its structure, and in a precisely similar way we have worked out the structure of acetyl. We represent the structure in the two cases thus:

```
  H H             O H
  | |             ‖ |
H-C-C-           -C-C-H
  | |               |
  H H               H
 Ethyl.          Acetyl.
```

Hence we conclude that the structure of a molecule of acetic ether should be represented as follows:

```
  H H     O H
  | |     ‖ |
H-C-C-O-C-C-H
  | |       |
  H H       H
  Acetic Ether.
```

Moreover, since we are led to the same result, whether we study the reactions by which the ether may be prepared or those by which it may be decomposed, we feel great confidence in our result.

If, now, we act on butyric acid, the isomer of acetic ether, with potassic hydrate, the same reagent as before, we obtain wholly different products. They are potassic butyrate and water; and here the knowledge of acids, bases, and salts, which we obtained at the last lecture, comes in to help us interpret the reaction. It must be simply as follows:

$$\left.\begin{matrix} \text{H-O-}C_4H_7O \\ \text{Butyric Acid.} \\ \text{K-O-H} \\ \text{Potassic Hydrate.} \end{matrix}\right\} = \left\{\begin{matrix} \text{K-O-}C_4H_7O \\ \text{Potassic Butyrate.} \\ \text{H-O-H} \\ \text{Water.} \end{matrix}\right.$$

Evidently, then, butyric acid, instead of containing the two radicals C_2H_5 and C_2H_3O, like acetic ether, contains the more complex radical C_4H_7O, and the simple radical H.

But, although the last reaction shows that butyric acid contains the radical C_4H_7O, it gives us no information in regard to the grouping of the atoms in the radical. Of course, we have sought to discover what the structure is, and the result of the investigation is most remarkable, for it appears that there are two different radicals having the same composition and corresponding to two distinct varieties of butyric acid, which differ in their odor, their boiling-point, and other qualities, and, further, various reactions show that the atoms of the radicals are arranged in the two acids as the following formulæ indicate:

```
                                                   H
                                                   |
      O   H   H   H                          O  H - C - H
      ‖   |   |   |                          ‖      |
H - O - C - C - C - C - H              H - O - C ——— C - H
          |   |   |                                 |
          H   H   H                          H - C - H
                                                    |
                                                    H
```

Normal Butyric Acid.
(Prepared from butter or by fermentation.)

Isobutyric Acid.
(A product of synthesis.)

There are, therefore, at least three substances having the composition $C_4H_8O_2$.

Now, by studying in a similar way the whole scheme of carbon compounds, and connecting by reactions the more complex with the simpler, it has been found possible, in a very large number of instances, to determine the manner in which the atoms are grouped in the respective molecules, and thus to show what the variations of structure are which determine the difference of qualities in these isomeric bodies. Moreover, having discovered how the atoms are grouped, it has been found possible, in many cases, to *reproduce* the compounds; and, more than this, chemists have frequently been led to the discovery of wholly new bodies, isomeric with old compounds, by studying the possible variations of the structural symbol. This last fact has such an important bearing on our subject, tending greatly to substantiate the general truth of our theory of molecular structure, that a few illustrations will be interesting. One of these we have already seen, for the isomeric modification of butyric acid, we have just been discussing, was foreseen by theory before it was discovered, and it is, therefore, an example in point, but there are many other cases of the kind which are equally remarkable.

Butyric acid is the fourth body in that series of volatile acids before mentioned (page 277), of which formic and acetic acids are the first and second members. It was then said that the molecules of these acids increase in weight by successive additions of CH_2 as we descend in the series, and it has been shown since (page 296), that the radical ethyl, $\begin{array}{c}H\ \ H\\ |\ \ \ |\\ -C-C-H\\ |\ \ \ |\\ H\ \ H\end{array}$, may be derived from methyl, $\begin{array}{c}H\\ |\\ -C-H\\ |\\ H\end{array}$, by replacing the terminal H by another methyl group. It is obvious that this process repeated on $\begin{array}{c}H\ \ H\\ |\ \ \ |\\ -C-C-H\\ |\ \ \ |\\ H\ \ H\end{array}$ would give $\begin{array}{c}H\ \ H\ \ H\\ |\ \ \ |\ \ \ |\\ -C-C-C-H\\ |\ \ \ |\ \ \ |\\ H\ \ H\ \ H\end{array}$, and that the result of successive replacements of the same kind would be a series of hydrocarbon radicals differing from each other by CH_2 like the volatile acids mentioned above. Furthermore, it is equally obvious that, theoretically at least, the same process might be applied to any compound containing a hydrocarbon radical; and you will not be surprised, therefore, to learn that there are many series of carbon compounds, between whose members we find this same common difference. Bodies so related are said to be the homologues of each other; and of these homologous series, so called, no one has been more carefully studied than that of the volatile acids, of which nineteen members are known.

Now, it is obvious that, as the hydrocarbon radical in the series of volatile acids increases in complexity, the possibilities of varying the atomic grouping increase also. Next to butyric acid, $C_4H_8O_2$, comes valeric acid, $C_5H_{10}O_2$, and, while we had only two butyric acids, we can have four valeric acids, whose molecular structure is indicated by the following symbols:

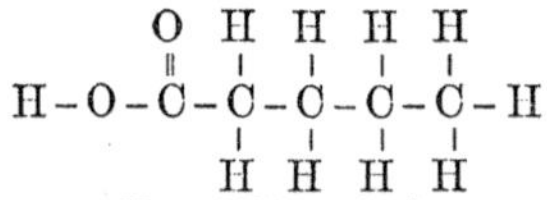

```
                                                  H
                                                  |
        O   H   H   H   H               O  H  H - C - H
        ‖   |   |   |   |               ‖  |      |
H - O - C - C - C - C - C - H   H - O - C - C ——— C - H
            |   |   |   |                  |      |
            H   H   H   H                  H  H - C - H
                                                  |
                                                  H
```

Normal Valeric Acid.

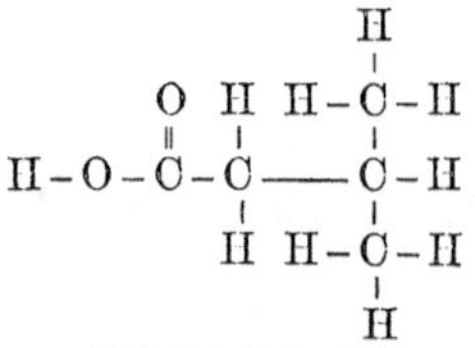

First Isovaleric Acid.

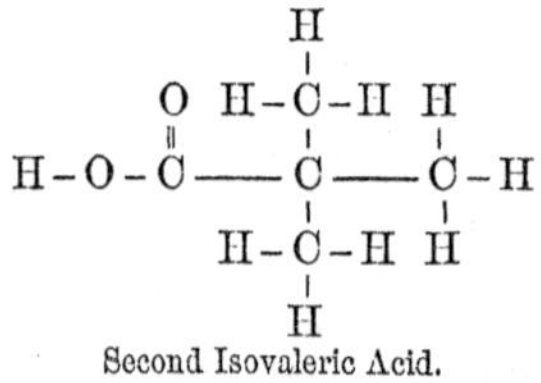

```
            H                           H
            |                           |
    O   H - C - H   H           O   H - C - H   H   H
    ‖       |       |           ‖       |       |   |
H - O - C ——— C ——— C - H   H - O - C ——— C ——— C - C - H
            |       |                   |       |   |
        H - C - H   H                   H       H   H
            |
            H
```

Second Isovaleric Acid.

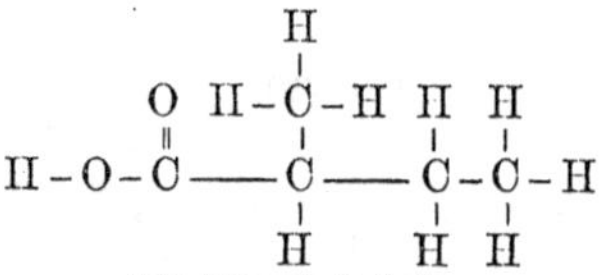

Third Isovaleric Acid.

Of these possible modifications of valeric acids, pointed out by theory, the first three have already been identified in the investigations to which the theory led, and the discovery of the fourth is probably only a question of time. Examples similar to this are already numerous and are rapidly multiplying, but I have only time to cite one other instance.

A compound called cyanic ether has long been known, and its symbol was always assumed to be—

$$(C_2H_5)\text{-}O\text{-}C{\equiv}N,$$

after the analogy of the other ethers; that is, it was assumed to contain the compound radicals, ethyl, C_2H_5, and cyanogen, CN, united through an atom of oxygen. But, as is obvious, we may, without changing the radical ethyl, group the other atoms thus:

$$(C_2H_5)\text{-}N{=}C{=}O,$$

and, on searching for this substance, an isomer of the supposed cyanic ether was actually obtained, and called cyanetholine. Very singularly, however, further investigation proved that the new compound was the real cyanic ether, and that the old one had the constitution represented by the last symbol. Evidently, then, we are not infallible; but the very mistake has been instructive; for, in detecting and correcting the error, we have the more clearly shown that our methods are trustworthy.

I hope I have been able to give some general notions of the manner in which we have obtained our knowledge of the grouping of the atoms in the compounds of carbon. More than this cannot be expected in a popular lecture; for, so interwoven is the web of evidence on which the conclusions are based, that, to enter into full details in regard to any one of the more complex compounds, would be wearisome, and the work is much better suited for the study than the lecture-room. Indeed, I fear that I have already imposed too great a burden on your patience; but, if you have followed me thus far, you will be interested in some of the results which we have reached, and which you are now prepared to understand. I must necessarily present these results as they have been formulated by our theory of atomic bonds; for, without the aid of these formulæ, we cannot either think or talk clearly about the subject.

The one characteristic of carbon on which the great complexity and variety of its compounds depend is, the power which its atoms possess of combining among themselves to an almost indefinite extent. As a rule, chemical combination takes place readily only between dissimilar atoms. It is true that we have met with many examples of the union of similar atoms, as in the molecules of several of the elementary gases, like—

$H-H$	$Cl-Cl$	$O=O$	$N\equiv N$
Hydrogen Gas.	Chlorine Gas.	Oxygen Gas.	Nitrogen Gas.

So, also, in the compounds—

$$\begin{array}{c} Cl \quad Cl \\ | \quad\;\; | \\ Cl-Fe-Fe-Cl \\ | \quad\;\; | \\ Cl \quad Cl \end{array} \quad \text{and} \quad \begin{array}{c} O \\ / \; \backslash \\ Al-Al \\ // \quad \backslash\backslash \\ O \qquad O \end{array}$$

Ferric Chloride. Aluminic Oxide.

and likewise in

$$Cl-Hg-Hg-Cl \quad \text{and} \quad \begin{array}{c} O \\ / \; \backslash \\ Cu-Cu \end{array}$$

Mercurous Chloride. Cuprous Oxide—

two atoms are united by a single bond, forming a binary group, which is the radical of the metallic compound. But, in all these cases, the power of combination is very limited, admitting the grouping together of only a very few atoms at the most, and generally of only two. The carbon-atoms, however, not only unite with each other in large numbers, but form groups of great stability, which, in organic compounds, take the place of the elementary radicals of the mineral kingdom. Let us begin, then, by constructing these radicals:

The carbon-atoms being quadrivalent, they may unite with each other either by one, two, three, or four bonds, and the larger the number of bonds which are thus closed, the less will evidently be the combining power of the resulting radical. Hence may arise radicals like those represented in the diagram on the previous page. It is evident that this table might be extended indefinitely, but the number of terms given is sufficient to illustrate the simple relation between the several radicals thus formed. Each group of carbon-atoms can have a maximum quantivalence of 2n + 2 (the letter n denoting the number of carbon-atoms in the group), and from this maximum the quantivalence may fall off by two bonds at a time until it is reduced to zero. Thus we have for the six-atom group a maximum of 14; but the same group may also have a quantivalence of 12, 10, 8, 6, 4, or 2.

The symbols, however, given in the table do not by

No. 1.

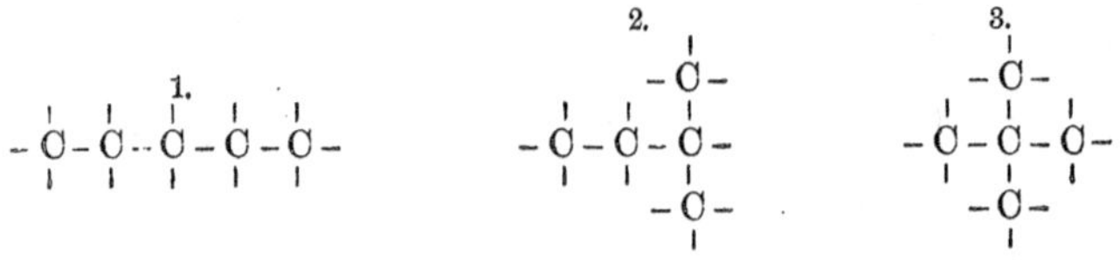

```
                                     2.                3.
                                      |                 |
          1.                        - C -             - C -
  |   |   |   |   |           |   |   |           |   |   |
- C - C - C - C - C -       - C - C - C -       - C - C - C -
  |   |   |   |   |           |   |   |           |   |   |
                                    - C -             - C -
                                      |                 |
```

No. 2.

```
          1.                        2.
  |   |   |   |             |   |   |   |
- C - C = C - C -           C = C - C - C -
  |           |             |       |   |

 3.                 4.
  |                \ /                     5.
- C -               C                       |
|  |               / \ / \           |   | / C -
C = C             C     C          - C - C   |
|  |               \ / \ /           |     \ C -
- C -               C                        |
  |                / \
```

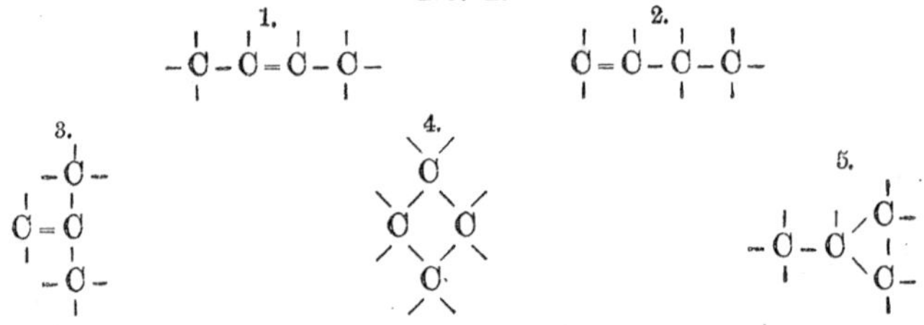

any means exhaust the possibilities of combination with the given number of carbon-atoms; for further variations may be obtained by changing the relative position of the atoms while retaining the same quantivalence. Thus, the radical $(C_5)^{xii}$ may be constructed in the several ways shown in diagram No. 1, and, although the several radicals thus obtained contain the same number of atoms, and have the same quantivalence, they are fundamentally different. The difference consists, not in the mere grouping of the letters on the page, which is purely arbitrary, but in the fact that, while in 1 no carbon-atom is united with more than two others, in 2, one of the atoms is united with three others, and, in 3, with four. As the number of atoms in the group increases, the number of possible variations must necessarily become very greatly augmented. Moreover, when some of the atoms are united by double bonds, a variation may be obtained by shifting the position of this double bond as well as by varying the position of the atoms with respect to each other. This is illustrated by diagram No. 2, which shows the possible forms of the group $(C_4)^{viii}$. It is unnecessary, however, to multiply illustrations; for it is evident that a great multitude of radicals may be obtained with even a very limited number of carbon-atoms, and to attempt to exhaust the possibilities would be an endless task. Some of my audience, however, may be interested to study the subject further, and I would, therefore, set them as a problem to find the number of possible combinations which can be made with a group of six carbon-atoms, having a quantivalence of twelve. Such investigations are not without their profit; for, although many of the possibilities may not be realized in Nature, yet the practice will give a clear idea of what is meant by an essentially different structure. It may hereafter appear that changes of position corresponding to the upper and lower, or the left and right hand sides of our diagram, constitute really essential variations of structure; but, although there are some facts looking in this direction, we do not as yet admit that any such differences are of importance, and we regard any two groups as the same when, by any change that does not alter the relative order of the atoms, or the number of bonds by which they are united, the two can be made to coincide thus:

```
        |                     |                    \ /  /
       -C-                   -C-                    C-C
  |  |  |                |  |  |                  \ /    \\
 -C-C-C- is the same as -C-C-C-, and            C      C- the same
  |  |  |                |  |  |                  / \    /
       -C-                   -C-                    C=C
        |                     |                     /  \
```

```
     \   /                                  \   /           \  \ /
      C-C                                    C=C             C-C
    //   \\                                \ /  \ /         //  \ /
as C       C,  but not the same as          C     C     or  C     C
  / \     / \                              / \  / \          \\  / \
     C-C                                     C=C              C-C
    / \ / \                                 /   \            /  / \
```

The radicals thus formed may be regarded as the skeletons of the organic compounds. These carbon-atoms, locked together like so many vertebræ, form the framework to which the other elementary atoms are fastened, and it is thus that the complex molecular structures, of which organized beings consist, are rendered possible; moreover, when we remember that, while the elementary substance carbon is a fixed solid, the three elementary substances, oxygen, hydrogen, and nitrogen, with which it is usually associated, are permanent gases, this analogy of the carbon-nucleus to the skeleton of the vertebrate animal becomes still more striking.

Having thus shown how the skeletons may be formed, let us next see how these dry bones may be clothed. In order to illustrate this point, I will simply take two of the numberless carbon-radicals, which are theoretically possible, and show how from them a set of familiar organic products can be derived. Let the two be the radicals represented in this diagram :

$$
\begin{array}{ccccc}
 & | & | & | & \\
- & C - & C - & C & - \\
 & | & | & | &
\end{array}
\qquad\qquad
\begin{array}{ccccc}
 & & | & & \\
 & & C & & \\
 & / & & \backslash\!\backslash & \\
-C & & & & C- \\
\| & & & & | \\
-C & & & & C- \\
 & \backslash & & /\!/ & \\
 & & C & & \\
 & & | & &
\end{array}
$$

To such carbon-skeletons a large number of different elementary atoms and compound radicals can be attached by various chemical processes ; but the number of those usually met with in organic compounds is very limited, and only the following will be considered in this connection, namely :

$H-$,	$-O-$,	$H-O-$,	$\begin{array}{l}H\\H\end{array}\!\!>N-$,	$\begin{array}{l}O\\O\end{array}\!\!\gg N-$
Hydrogen.	Oxygen.	Hydroxyl.	Amidogen.	Nitryl.

Indeed, by doubling this number, we could obtain the materials for constructing nearly the whole scheme of modern organic chemistry.

Beginning, then, with the nucleus $\begin{array}{c} \;\;|\quad\;|\quad\;| \\ -C-C-C- \\ \;\;|\quad\;|\quad\;| \end{array}$, let us, in the first place, satisfy all the open bonds with hydrogen-atoms. The result is—

$$
\begin{array}{ccccccc}
 & H & & H & & H & \\
 & | & & | & & | & \\
H - & C & - & C & - & C & - H \\
 & | & & | & & | & \\
 & H & & H & & H &
\end{array}
$$

Propylic Hydride.

a combustible gas, which is found mixed with numerous other compounds of the same class in our petroleum-wells. Propyl hydride is the third in a series of homologous compounds, of which no less than nine have been identified in our Pennsylvania petroleums.

Methylic hydride	$C\ H_4$	Gas.
Ethylic hydride	C_2H_6	"
Propylic hydride	C_3H_8	"
Butylic hydride	C_4H_{10}	32°
Amylic hydride	C_5H_{12}	86°
Hexylic hydride	C_6H_{14}	142°
Heptylic hydride	C_7H_{16}	194°
Octylic hydride	C_8H_{18}	247°
Nonylic hydride	C_9H_{20}	303°

The diagram, above, gives their names and boiling-points. Our common kerosene is chiefly a mixture of hexylic and heptylic hydride, and the light naphthas a mixture of amylic and hexylic hydrides. Notice here, again, the common difference, CH_2, between the symbols of any two consecutive members of this series of hydrocarbons.

If, next, we substitute an atom of oxygen for two of the hydrogen-atoms which, in propylic hydride, are united to either of the terminal atoms of the carbon-nucleus, we obtain a compound called propylic aldehyde. This is a member of another series of homologues, parallel to the last, and of which nearly as many members are known. The aldehydes, as these bodies are all called, have very striking and characteristic qualities; and these qualities may be, to a great extent, traced to their peculiar molecular structure. If we only make so small a change as to transfer the oxygen-atom from the terminal to one of the central atoms of the carbon-nucleus, we obtain a class of compounds which, though isomeric with the aldehydes, have wholly different qualities, and are called ketones. The ketone isomeric with propylic aldehyde is called acetone:

```
   H   H   O                H   O   H
   |   |   ||               |   ||  |
H - C - C - C - H        H - C - C - C - H
   |   |                    |       |
   H   H                    H       H
Propylic Aldehyde.           Acetone.
```

Going back again to the hydrocarbon, C_3H_8, and replacing either of the terminal hydrogen-atoms by the radical hydroxyl (-O-H), we obtain one of a very important class of compounds, called alcohols.

$$\begin{array}{c} H\ \ H\ \ H \\ |\ \ \ |\ \ \ | \\ H-C-C-C-H \\ |\ \ \ |\ \ \ | \\ H\ \ H\ \ H \end{array} \quad \text{gives} \quad \begin{array}{c} H\ \ H\ \ H \\ |\ \ \ |\ \ \ | \\ H-C-C-C-O-H \\ |\ \ \ |\ \ \ | \\ H\ \ H\ \ H \end{array}$$

Propylic Hydride gives Propylic Alcohol.

Propylic alcohol is the third member of still another series of homologous compounds, of which our common alcohol is the second member.

Normal Alcohols.

Methylic alcohol (wood-spirit)	$C\ H_3$-O-H
Ethylic alcohol (common alcohol)	$C_2\ H_5$-O-H
Propylic alcohol	$C_3\ H_7$-O-H
Butylic alcohol	$C_4\ H_9$-O-H
Amylic alcohol (fusel-oil)	$C_5\ H_{11}$-O-H
Hexylic alcohol	$C_6\ H_{13}$-O-H
Heptylic alcohol	$C_7\ H_{15}$-O-H
Octylic alcohol	$C_8\ H_{17}$-O-H

The structure of the alcohol may obviously be varied, like that of the aldehyde, by transferring the hydroxyl from the terminal to one of the central atoms of the carbon-nucleus; but we thus, as before, obtain a wholly new set of substances, which, although resembling the normal alcohols in many respects, differ from them in important particulars. There is, for example, an isopropylic alcohol, which is isomeric with the normal propylic alcohol, and, like it, resembles externally common alcohol. But the pseudo-alcohol, as we call it, boils at 85° Cent., while the normal alcohol boils at 97°, and, when acted on by chemical agents, yields wholly different products:

```
    H  H  H                      H  H  H
    |  |  |                      |  |  |
H - C - C - C - O - H        H - C - C - C - H
    |  |  |                      |  |  |
    H  H  H                      H  O  H
                                    |
                                    H
```

Propylic Alcohol. Isopropylic Alcohol.

Continuing, now, this process of clothing the carbon-skeleton, let us, in the next place, substitute for two of the hydrogen-atoms of the normal alcohol an atom of oxygen, selecting for replacement the two hydrogen-atoms which are connected with that terminal carbon-atom to which the hydroxyl is united:

```
    H  H  H                      H  H  O
    |  |  |                      |  |  ||
H - C - C - C - O - H        H - C - C - C - O - H
    |  |  |                      |  |
    H  H  H                      H  H
```

Propylic Alcohol gives Propionic Acid.

Now, propionic acid is the third member of that homologous series of volatile acids of which a partial list has already been given (page 277), and of two of whose members the possible variations of structure have already been discussed (pages 298 and 300).

Again, we may substitute in propionic acid a second oxygen atom for two of the remaining atoms of hydrogen, and we thus obtain a liquid body called pyruvic acid, a perfectly definite substance, although one with which I can give you no familiar associations:

```
    H  H  O                      H  O  O
    |  |  ||                     |  || ||
H - C - C - C - O - H        H - C - C - C - O - H
    |  |                         |
    H  H                         H
```

Propionic Acid. Pyruvic Acid.

The acids and alcohols we have thus far formed around our three-atom carbon-nucleus have been all monatomic. The atomicity of a compound, you remember, is determined by the number of atoms of hydrogen which are easily replaced by metathesis, and only those atoms of hydrogen can be so replaced which are united to the carbon-nucleus through an atom of oxygen. Hence, with one hydroxyl group we can only produce monatomic compounds. Use two hydroxyl groups, and we can form around the same skeleton a number of diatomic compounds. The following are a few examples. After what has been said, the symbols require no detailed description; but it must be remembered that the grouping is no play of fancy, and that a good reason can be given for the position of every letter:

```
      H   H   H
      |   |   |
H - O - C - C - C - O - H
      |   |   |
      H   H   H
```

(Not yet discovered.)
Normal Propyl Glycol.

```
      H   H   H
      |   |   |
H - O - C - C - C - H
      |   |   |
      H   O   H
          |
          H
```

Propyl Glycol.

```
      O   H   H
      ‖   |   |
H - O - C - C - C - O - H
          |   |
          H   H
```

Normal or Paralactic Acid.

```
      O   H   H
      ‖   |   |
H - O - C - C - C - H
          |   |
          O   H
          |
          H
```

Common Lactic Acid.

Attach to the nucleus three hydroxyl groups, and there result triatomic compounds, among which is a very familiar substance:

```
      H   H   H
      |   |   |
H - O - C - C - C - O - H
      |   |   |
      H   O   H
          |
          H
```

Glycerine.

```
      O   H   H
      ‖   |   |
H - O - C - C - C - O - H
          |   |
          O   H
          |
          H
```

Glyceric Acid.

```
      O   H   O
      ‖   |   ‖
H - O - C - C - C - O - H
          |
          O
          |
          H
```

Tartronic Acid.

Lastly, replace the three terminal hydrogen-atoms of glycerine by nitryl (NO_2), and we meet again an old acquaintance:

$$\begin{array}{ccccccccccc}
O & & & & H & & H & & H & & & & O \\
\| & & & & | & & | & & | & & & & \| \\
N & - & O & - & C & - & C & - & C & - & O & - & N \\
\| & & & & | & & | & & | & & & & \| \\
O & & & & H & & O & & H & & & & O \\
 & & & & & & | & & & & & & \\
 & & & & & O = & N & = O & & & & &
\end{array}$$

Nitro-glycerine.

I think that this last symbol will not now appear to you so strange as when I first called your attention to it a few lectures back. It is true that I have not actually proved that this grouping of the letters represents the structure of the nitro-glycerine molecule, but I have led you to a point where you are prepared to accept it as a definite result of investigation, and can feel assured that the proofs await your examination in the due course of your study. You can now understand more clearly than before how it is that, by the structure of the molecule, the oxygen-atoms are kept apart from the atoms of carbon and hydrogen for which the fire-element has such a strong affinity, and how these atoms rush into more stable combinations when the delicate balance of forces, on which the structure depends, is disturbed.

You have now seen what a number of distinct compounds can be obtained by attaching to one of the very simplest of the carbon-nuclei atoms of hydrogen and oxygen alone. Almost every commutation we could make with these few atoms is actually realized in a definite substance. Of course, with the names of many of these bodies you have no association. You must accept the assurance that they stand for definite substances, and that our symbols represent the results of careful investigation, and, knowing this, you can gain some conception of the knowledge we have acquired of the structure of this class of compounds; and, when you add to this that, in many of these cases, the theory has gone before discovery, and, by suggesting possible commutations of the atoms, has prefigured compounds which were subsequently obtained, you must admit that, rude and unreal as our representations of molecular structure may be, they have a positive value, both as means of classifying facts and as aids to new discoveries.

Lastly, let us turn our attention to the second of the two carbon-skeletons, whose dry bones we proposed to clothe with the features of definite compounds. The group of bodies whose molecules contain, as we assume, this nucleus (Fig. 32), has been very fully investigated by Professor Kekulé, of Bonn, and to him we owe the theory of their structure which our diagram represents. It may appear superfluous for me to repeat that, in such diagrams, the only essential points are the relative order of the atoms and the number of the bonds; but the hexagonal shape in which we find it convenient to represent on our page the structure of this nucleus suggests the idea of definite form so forcibly, that additional caution may be needed to avoid misconstruction.

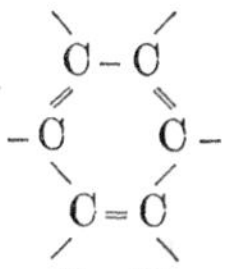

```
 \   /
  C - C
 //    \\
-C      C-
 \     /
  C = C
 /     \
```

Fig. 32.

The bodies with which we are now to deal are, for the most part, products either already existing in coal-tar, or which may be obtained from it by various chemical processes. Among them are those gorgeous aniline dyes which, within a comparatively few years, have added so much to the elegances of common life. From a very large number of compounds, I can only select a few examples. Still, I shall not restrict the selection to compounds whose molecules contain only six carbon-atoms, but I shall endeavor to show that molecules of extreme complexity can be built up either by the addition of hydrocarbon radicals to the nucleus represented in Fig. 32, or by the coaleascing of two or more of these nuclei into one. As I have not time to enter into details, the symbols must, to a great extent, be allowed to speak for themselves.

Coal-tar is a mixture of a very large number of substances whose boiling-points vary from 80° Cent. upward. When the tar is distilled, and the distillate rectified, the more volatile product obtained is chiefly a mixture of two hydrocarbons—benzol and toluol. This mixture, the commercial benzol, is used in large quantities for the preparation of the aniline dyes:

```
      H   H                        H   H
      |   |                        |   |
      C - C                    H   C - C
     //    \\                  |  //    \\
H - C        C - H         H - C - C        C - H
     \      /                  |   \      /
      C = C                    H    C = C
      |   |                         |   |
      H   H                         H   H
     Benzol.                        Toluol.
```

When benzol and toluol are treated with strong nitric acid the products are:

```
      H   H                          H   H
      |   |                          |   |
      C - C    O                 H   C - C    O
     //    \\  ||                |  //    \\  ||
H - C        C - N           H - C - C        C - N
     \      /  ||                |   \      /  ||
      C = C    O                 H    C = C    O
      |   |                           |   |
      H   H                           H   H
  Nitrobenzol.          or        Nitrotoluol.
```

When nitrobenzol and nitrotoluol are acted on by nascent hydrogen (in the arts a mixture of iron-filings and acetic acid is used), we obtain:

```
     H  H                              H  H
     |  |                              |  |
     C - C   H                     H   C - C   H
    //    \\  |                     |  //    \\  |
H - C      C - N                H - C - C      C - N
    \     /   |                     |  \     /   |
     C = C   H                     H   C = C   H
     |  |                              |  |
     H  H                              H  H
  Aniline,              or           Toluidine.
```

When the mixture of aniline and toluidine, obtained in the arts from commercial benzol, is treated with various oxidizing agents, we obtain salts of

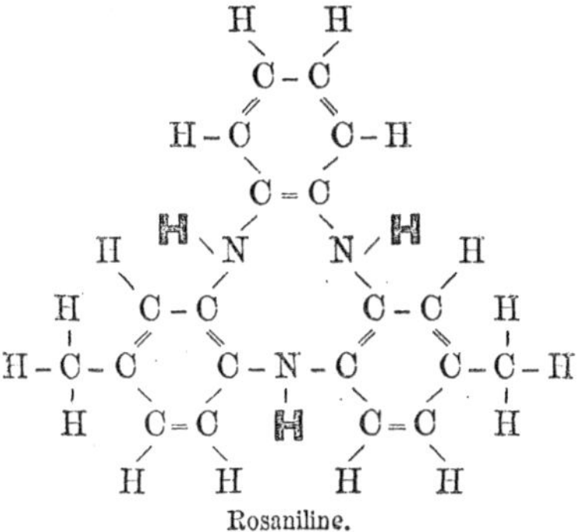

Rosaniline.

Rosaniline is a base like ammonia. As I have before stated, when the molecule NH_3 unites with acids to form salts, the quantivalence of the nitrogen-atom appears to be increased by two bonds which bind the atoms of the acid molecules (*see* page 238). So, when rosaniline combines with acids, the atoms of the acid molecule join to one or the other of the nitrogen-atoms in the complex molecule of this base. Moreover, as there are three of these nitrogen-atoms in the molecule of rosaniline, it can bind either one, two, or three molecules of acid; for example, it can unite either with HCl, with 2HCl, or with 3HCl. Thus, there may be formed three classes of salts, and those which contain the smallest amount of acid are used in the arts as coloring agents. These salts, when crystallized, have a very brilliant beetle-like lustre, and yield beautiful rose-red solutions. They possess, moreover, a most wonderful coloring power.

Taking only a few crystals (one grain in weight) of the hydrochlorate of rosaniline, called fuchsine in commerce, and, first rubbing them up in a mortar with some alcohol, I will pour the concentrated solution into a large glass jar, holding two gallons of water, and you see that this very small quantity of dye shows a brilliant red color even when diffused through the large body of liquid. By combining the base with different acids we obtain only slight variations of tint, but very marked alterations of color can be produced in another way.

By recurring to the symbol of rosaniline, it will be seen that there are three hydrogen-atoms directly united to the three atoms of nitrogen which the radical contains. Now, it is possible to replace either one, two, or all three of these hydrogen-atoms by various hydrocarbon radicals; such as—

$$\underset{\text{Methyl,}}{-CH_3} \qquad \underset{\text{Ethyl,}}{-C_2H_5} \quad \text{or} \quad \underset{\text{Phenyl.}}{-C_6H_5};$$

and we thus obtain other bases whose salts are violet or blue—the blue tint increasing with the degree of replacement. I have in these five jars solutions of some of these salts, the aniline violets and blues of commerce, and they will illustrate to you the gradations of color we can obtain by the replacements I have described.

Among the less volatile products of the distillation of coal-tar is the compound called phenol or carbolic acid, which is so much used as an antiseptic agent. Here is its symbol and also the symbol of another compound which has recently acquired great theoretical importance, but which, although closely allied to phenol, is derived from a wholly different source:

```
     H  H                     H  H
     |  |                     |  |
     C - C                    C - C
    //    \\                 //    \\
H - C      C - O - H    H - C      C - H
    \      /                 \      /
     C = C                    C = C
     |   |                    |   |
     H   H                    O - O
    Phenol.                  Quinone.
```

One of the least volatile products obtained in the distillation of coal-tar is a hydrocarbon called naphthaline, whose molecule appears to be formed by the coalescing of two molecules of benzol. This body yields a very large number of derivatives having the same general structure, some of which have such a deep color that they can be used as dyes:

```
          H     H
          |     |
          C     C
        //  \  /  \\
   H - C     C     C - H
       |     ||    |
   H - C     C     C - H
        \\  /  \  //
          C     C
          |     |
          H     H
         Naphthaline.
```

Associated with naphthaline in coal-tar is a still less volatile hydrocarbon, called anthracene, which may be regarded as formed by the coalescing of three molecules of benzol:

```
         H     H
         |     |
         C     C     H
       //  \ /   \\ /
  H - C     C     C     H
      |     ||    |    /
  H - C     C     C - C
       \\  / \  //     \\
         C     C         C - H
         |      \       /
         H       C  -  C
                 |     |
                 H     H
```

Anthracene.

Lastly, from anthracene has been derived the following product :[1]

```
        H
         \
          O     H
           \    |
            C   C     H
          // \ / \\  /
  O - C      C     C     H
  |   |      ||    |    /
  O - C      C     C - C
       \\   / \  //     \\
         C      C         C - H
        /        \       /
       O          C  =  C
      /           |     |
     H            H     H
```

Anthraquinonic Acid (Alizarine).

This brings us to one of the latest and most noteworthy results of our science. Alizarine is the coloring principle of the madder-root, which has long been the chief dyestuff used in printing calicoes. But, although the mordanted cloth extracts from a decoction of the root the coloring material in a condition of great purity, yet it has been found exceedingly difficult to isolate the alizarine. For this reason, although the subject had been most carefully investigated, there was for many years a question in regard to the exact com-

[1] For further details see "Principles of Chemical Philosophy," by Josiah P. Cooke, Jr., published by John Allyn, Boston, third edition, 1874.

position of the substance. A short time since, Graebe, a German chemist, in investigating a class of compounds called the quinones,[1] determined incidentally the molecular structure of a body closely resembling alizarine, which had been discovered several years before. This body was derived from naphthaline, and, like many similar derivatives, was reduced back to naphthaline when heated with zinc-dust. This circumstance led the chemist to heat also madder alizarine with zinc-dust, when, to his surprise, he obtained anthracene. Of course, the inference was at once drawn that alizarine must have the same relation to anthracene that the allied coloring-matter bore to naphthaline, and, more than this, it was also inferred that the same chemical processes which produced the coloring-matter from naphthaline, when applied to anthracene, would yield alizarine. The result fully answered these expectations, and now alizarine is manufactured on a large scale from the anthracene obtained from coal-tar.

Here are two pieces of cloth, one printed with madder and the other with artificial alizarine, and the most expert calico-printer could not distinguish between them.

This certainly is a most remarkable achievement. A highly-complex organic product has been actually constructed by following out the indications of its molecular structure, which the study of its reaction, and those of allied compounds, had furnished. It is a result that all can appreciate, and which the world will accept as the most trustworthy credential that the molecular

[1] The name quinone is applied to a class of bodies whose molecules contain two atoms of oxygen united to a carbon-nucleus in the peculiar way shown in the symbol of the typical compound of the class, given above (page 317).

theory of chemistry could offer. The circumstance that this substance is the important madder-dye, and that the new process has a great commercial value, of course, really adds nothing to the force of the evidence in favor of the theory. To the scientific mind the evidence of any one of hundreds of substances which have been constructed in a similar way, but of which the world at large has never heard, is equally conclusive. Still, we have great reason to rejoice that this is one of the few instances where purely theoretical study has been unexpectedly crowned with great practical results. Let us accept the gift with gratitude, and pay due honor to those through whose exertions it has been received. Let us remember, however, that it came as a free gift, and that the result was achieved by men who, with single-hearted zeal, worked solely to extend knowledge. Forget not, then, to encourage those who are devoting their lives to the same noble service, and have the manly courage to sow the seed whose harvest they can never hope to reap. Honor those who seek Knowledge for her own sake, and remember they are the great heroes of the world, who work in faith, and leave the result with God!

INDEX.

THE numbers of this index refer to pages. Attention is called to the lists of experiments, graphic symbols, reactions, and tables given under these several headings.

THE END.

www.ingramcontent.com/pod-product-compliance
Lightning Source LLC
LaVergne TN
LVHW020213110826
845151LV00003B/700

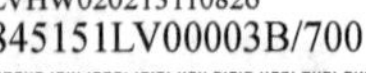